Reviews and critical articles covering the entire field of normal anatomy (cytology, histology, cyto- and histochemistry, electron microscopy, macroscopy, experimental morphology and embryology and comparative anatomy) are published in Advances in Anatomy, Embryology and Cell Biology. Papers dealing with anthropology and clinical morphology that aim to encourage cooperation between anatomy and related disciplines will also be accepted. Papers are normally commissioned. Originalpapers and communications may be submitted and will be considered for publication provided they meet the requirements of a review article and thus fit into the scope of "Advances". English language is preferred.

It is a fundamental condition that submitted manuscripts have not been and willnot simultaneously be submitted or published elsewhere. With the acceptance of a manuscript for publication, the publisher acquires full and exclusive copyright for all languages and countries.

Twenty-five copies of each paper are supplied free of charge.

Manuscripts should be addressed to

Co-ordinating Editor

Prof. Dr. H.-W. KORF, Zentrum der Morphologie, Universität Frankfurt, Theodor-Stern Kai 7,
60595 Frankfurt/Main, Germany
e-mail: korf@em.uni-frankfurt.de

Editors

214

Advances in Anatomy, Embryology and Cell Biology

For further volumes:
http://www.springer.com/series/102

Friedemann Kiefer •
Stefan Schulte-Merker
Editors

Developmental Aspects of the Lymphatic Vascular System

with 27 Figures

Editors
Friedemann Kiefer
MPI for Molecular Biomedicine
Münster
Germany

Stefan Schulte-Merker
Hubrecht Institute Development Biology
Utrecht
The Netherlands

ISSN 0301-5556 ISSN 2192-7065 (electronic)
ISBN 978-3-7091-1645-6 ISBN 978-3-7091-1646-3 (eBook)
DOI 10.1007/978-3-7091-1646-3
Springer Wien Heidelberg New York Dordrecht London

Library of Congress Control Number: 2013954874

Printed on acid-free paper

Springer is part of Springer Science+Business Media (www.springer.com)

Abbreviations

2P-LSM	Two-photon laser scanning microscopy
3D	Three dimensional
aISV	Arterial intersegmental vessel
AD	Autosomal dominant
APC	Antigen-presenting cell
AR	Autosomal recessive
BAC	Bacterial artificial chromosome
BALT	Bronchus-associated lymphoid tissue
BEC	Blood endothelial cell
BM	Basement membrane
BMDMC	Bone marrow-derived myeloid cell
Brg1	ATP-dependent chromatin-remodeling enzyme brahma-related gene 1
BV	Blood vessel
CCBE-1	Collagen and calcium-binding EGF domain-1
CCL19	C-C motif chemokine 19
CCL21	C-C motif chemokine 21
CCR7	C-C chemokine receptor type 7
CCV	Common cardinal vein
CD36	Cluster of differentiation 36, thrombospondin receptor
CLEC	C-type lectin domain family member
CnB1	Calcineurin Ca^{2+}-binding regulatory subunit B1
COUP-TFII	Chicken ovalbumin upstream promoter-transcription factor II
CsA	Cyclosporin A
CT	Computer tomography
CV	Cardinal vein
CVM	Congenital vascular malformation
Cx37	Connexin 37
Cx43	Connexin 43
Cxcl12	C-X-C motif chemokine 12, also known as stromal cell-derived factor 1 (SDF-1)

Cxcl13	C-X-C motif chemokine 13, also known as B lymphocyte chemoattractant (BLC)
CXCR4	C-X-C chemokine receptor type 4, receptor for chemokine Cxcl12
CXCR5	C-X-C chemokine receptor type 5, receptor for chemokine Cxcl13
dpf	Days post fertilization
dsRED	*Discosoma* sp. red fluorescent protein
DA	Dorsal aorta
DC	Dendritic cell
DLAV	Dorsal longitudinal anastomotic vessel
DLL4	Delta-like ligand 4
EC	Endothelial cell
ECM	Extracellular matrix
EM-CCD	Electron multiplying charge-coupled device (sensitive digital camera chip)
EMCN	Endomucin, endothelial sialomucin
ERM	Ezrin-Radixin-Moesin family actin-binding protein
ESAM	Endothelial cell-selective adhesion molecule
ET	Extratruncular
E(x)	Day x of embryonic development
FcγR	Receptor for Fc portion of IgG
FITC	Fluorescein isothiocyanate
FN	Fibronectin
Foxc2	Forkhead box protein C2
FP	Fluorescent protein
GAP	GTPase-activating/accelerating protein
Gata2	GATA-binding protein 2
GFP	Green fluorescent protein
GPVI	Glycoprotein VI receptor for collagen
hpf	Hours post fertilization
H&E stain	Hematoxylin/eosin staining
HB-EGF	Heparin-binding epidermal growth factor
Hey1/2	Hairy/enhancer-of-split related with YRPW motif protein 1
HLM	Hemolymphatic malformation
HSPG	Heparan sulfate proteoglycan
HUVEC	Human umbilical vein endothelial cell
iLEC	Initial lymphatic endothelial cell, migrating CV-derived lymphatic endothelial progenitor of mesenchymal appearance
ICAM-1	Intercellular adhesion molecule 1
ICG	Indocyanine green
IJV	Internal jugular vein
IL-1β	Interleukin 1 beta
INFγ	Interferon gamma
IP3	Inositol-1,4,5-trisphosphate
IPAS	Inhibitory PAS (Per, Arnt, Sim) domain protein

IR	Infrared
ISA	Intersegmental artery
ISV	Intersegmental vein/intersegmental vessel
ISLV	Intersegmental lymphatic vessel
ISSVA	International Society for the Study of Vascular Anomalies
Itga9	Integrin $\alpha9$
ITAM	Immunoreceptor tyrosine-based activation motif
Jag1	Jagged 1
JAM-A (JAM1)	Junctional adhesion molecule A/1
JLS	Jugular lymph sac
KO	"Knockout" gene deletion by homologous recombination
LD	Lymphedema-distichiasis
LEC	Lymphatic endothelial cell
LM	Lymphatic malformation
LN	Lymph node
LPS	Lipopolysaccharide
LSD	Lysosomal storage disease
LSI	Laser speckle imaging
$LT\alpha_1\beta_2$	Lymphotoxin-$\alpha_1\beta_2$
$LT\beta$-R	Lymphotoxin receptor-β
LTi	Lymphoid tissue inducer cells
LTo	Lymphoid tissue organizer cells
LV	Lymphatic vessel
LVC	Lymphatic valve forming cell
LYVE-1	Lymphatic vessel endothelial hyaluronan receptor 1
miR	microRNA
mOrange 2	(Photostabilized) monomeric orange fluorescent protein 2
MAPK/ERK	Mitogen-activated protein kinase/extracellular signal-regulated kinase
MHC	Major histocompatibility complex
MHz	Megahertz
MRI	Magnetic resonance imaging
MSC	Mesenchymal stem cell
MSM	Medullary sinus macrophages
$NF\kappa B$	Nuclear factor kappa-light-chain-enhancer of activated B cells
NFATc1	Nuclear factor of activated T-cells, cytoplasmic 1
NIR	Near infrared
Nrp1/2	Neuropilin-1/2
OCT	Optical coherence tomography
OPO	Optical parametric oscillator
OPT	Optical projection tomography
ORF	Open reading frame
pTD	Primordial thoracic duct

PAA Pharyngeal arch artery
PAM Photoacoustic microscopy
PAT Photoacoustic tomography
PCV Posterior cardinal vein
PDGF Platelet-derived growth factor
PDPN Podoplanin
PECAM-1 Platelet endothelial cell adhesion molecule 1
PET Positron emission tomography
PF4 Platelet factor 4, a small chemokine of the CXC family
PG Proteoglycan
PIM Light sheet or planar illumination microscopy
PL Parachordal lymphangioblast
PL Pulmonary lymphangiectasia
PLCγ Phosphoinositide-phospholipase C gamma
PLLV Peripheral longitudinal lymphatic vessel
PLXN Plexin
Prox1 Prospero homeobox protein 1
P(x) Postnatal day x
RAF1 v-*raf-1* (rapidly accelerated fibrosarcoma/rat fibrosarcoma) murine leukemia viral oncogene homolog 1
RAR Retinoic acid receptors
RbpJ Recombination signal binding protein for immunoglobulin kappa J region
RFP Red fluorescent protein
RXR Retinoid X receptors
sLEC Superficial lymphatic endothelial cell
sVP Superficial venous plexus
SCV Subclavian vein
SEMA Semaphorin/semaphorin domain
SFK Src family kinase
SHG Second harmonic generation
SLP-76 SH2 domain-containing leukocyte protein of 76 kDa
SMC Smooth muscle cell
SOX 18 SRY (sex determining region Y)-box 18
SPECT Single-photon emission computed tomography
SSM Subcapsular sinus macrophages
Syk Spleen tyrosine kinase
tdTomato Tandem-dimer tomato fluorescent protein
T Truncular
Tbx1 T-box 1
TGFβ Transforming growth factor beta
TLO Tertiary lymphoid organs
TNFα Tumor necrosis factor alpha
TSP-1/2 Thrombospondin-1/2

vISV	Venous intersegmental vessel
VE-Cadherin	Vascular endothelial cadherin
VEC	Venous endothelial cell
VEGF-C/D	Vascular endothelial growth factor C/D
VEGFR-1/2/3	Vascular endothelial growth factor receptor 1/2/3
VM	Venous malformation
WBDD	Winter-Baraitser Dysmorphology Database
XL	X-linked recessive
ZAP-70	Zeta chain associated protein of 70 kDa
ZO-1	Zonula occludens 1

Contents

Chapter 1
Introductory Remarks

Friedemann Kiefer and Stefan Schulte-Merker

Abstract This introductory section briefly highlights the subsequent chapters in the context of recent findings and open questions in lymphatic vessel biology. It aims to provide a quick overview and orientation in the contents of this monograph collection.

1.1 Introductory Remarks

After their first mention by Erasistratus of Chios (304–250 BC), lacteal lymphatic vessels were rediscovered in the late sixteenth century by Gaspare Aselli (1581–1626) followed by the first description of the thoracic duct in the mid-seventeenth century by Thomas Bartholin (1616–1680). Although the principal function of the lymphatic vascular system to return fluids into the blood circulation was recognized nearly in parallel with the understanding of the blood circulatory system, the lymphatic system remained the "pale brother" and severely under-investigated until the last years of the previous century. Shortly before the turn of the millennium, the identification of positively identifying molecular markers, apparently specific growth factors, and the generation of mouse mutants with selective defects in the lymphatic vascular system sparked a renaissance in lymphatic vessel research (Oliver and Detmar 2002).

F. Kiefer (✉)
Mammalian Cell Signaling Laboratory, MPI for Molecular Biomedicine,
Röntgenstrasse 20, 48149 Münster, Germany
e-mail: fkiefer@gwdg.de

S. Schulte-Merker
Developmental Biology, Hubrecht Institute, KNAW and UMC Utrecht, Uppsalalaan 8, 3584 Utrecht, The Netherlands

Experimental Zoology Group, University of Wageningen, Wageningen, The Netherlands
e-mail: s.schulte@hubrecht.eu

F. Kiefer and S. Schulte-Merker (eds.), *Developmental Aspects of the Lymphatic Vascular System*, Advances in Anatomy, Embryology and Cell Biology 214,
DOI 10.1007/978-3-7091-1646-3_1,

The developmental origin of the lymphatic endothelium had remained unclear until in the first years of the twentieth century, when Sabine proposed after extensive ink injection experiments in pig embryos that the jugular lymph sacs arose from the anterior cardinal vein (Sabin 1902, 1916) and also other first lymphatic structures were of venous descent. The lymphatic vasculature was then suggested to develop by sprouting lymphangiogenesis and remodelling from the chain of first lumenized lymphatic structures established in the embryo. By now, analysis of targeted gene deletions and genetic lineage tracing studies in mice have confirmed Sabine's model (Srinivasan et al. 2007).

Lymphatic endothelial cell (LEC) progenitors are specified in the cardinal vein (CV) of the embryo and their emergence and further differentiation is governed by a transcriptional network, at the center of which act the transcription factors CoupTFII, Sox18, and Prox1. In particular Prox1 has been shown to be indispensible but also sufficient for the acquisition and maintenance of lymphatic identity. Establishment and action of the LEC transcription factor network have been intensely studied over the past decade and are reviewed in Chap. 2. Committed Prox1-expressing LEC progenitors shortly after their appearance leave the CV and establish the first lumenized lymphatic structures that are collectively referred to as primary lymph sacs. The precise mechanism of this process has been debated over some time and now very recently been solved as delamination and migration of LEC progenitors with a mesenchymal appearance. The respective studies are reviewed in detail in Chaps. 2 and 13.

At least in the mouse shortly after midgestation, the first, now lumenized, lymphatic vessels become functional and start to drain accumulating tissue fluid. Failure of functional lymphatic vessel formation has been shown to result in edema in different settings. Remarkably, mechanical stress exerted by increasing tissue fluid accumulation directly exerts a strong regulatory influence on LECs, which respond to the stretching force resulting from increased fluid pressure by proliferation. The molecular details which so elegantly couple mechanical stress and lymphangiogenesis have recently started to become uncovered and are presented in Chap. 3.

During functional differentiation of lymphatic vessels major differences in the endothelial cell junctional structures, basement membrane composition, and mural cell coverage are established that distinguish mature lymphatic from blood vessels. Between the mature LECs of lymphatic capillaries also called initial lymphatics, cell junctions display a characteristic discontinuous distribution of the adhesion molecule VE-cadherin. These specialized interrupted cell junctions are referred to as buttons. Button-like junctions are a functional specialization that is only acquired after birth and remarkably is reversible during inflammation. Chapter 4 describes the development and remodelling of this functionally important structure in detail.

The extracellular matrix (ECM) provides structural support to lymphatic endothelial cells, but is also distinctly designed to act as an integral part of lymphatic vessel functional architecture. As such, the ECM directs and modulates important aspects of lymphangiogenesis and ECM composition as well as ECM cell surface receptors of lymphatic endothelium are discussed in Chap. 5.

During the differentiation of lymphatic vessels, one important final step is the formation of valves in collecting lymphatic vessels. The development of bicuspid valves is indispensible for the establishment and maintenance of a directional lymph flow and valve failure is intricately associated with lymphedema formation. Valve development is initiated after the onset of lymphatic drainage and Chap. 6 discusses intriguing findings that suggest flow and mechanotransduction may cooperate to trigger and sustain a genetic program, which controls the formation and regulation of lymphatic valves.

An important functional aspect of the lymphatic system is the provision of a highway for antigen-presenting cells. The development of lymph nodes and lymphatic vessels is intimately interconnected and Chap. 7 outlines the cellular and molecular players that provide spatial information and orchestrate the coordinate building of these structures.

Yet, lymphatic vessels do not only provide a trafficking route for hematopoietic cells, but they also functionally interact with the hematopoietic system, an aspect that is only starting to be understood. One of the most enigmatic puzzles is the interaction of platelets and lymphatic vessels, which is described in Chap. 8. Platelets fulfil an essential role during the separation of the blood and lymphatic vascular systems, but are also indispensible for the maintenance of lymphatic vessel integrity.

In Chap. 9, an additional aspect of interaction between lymphatic vessels and the hematopoietic system is addressed. It becomes increasingly clear that lymphatic vessels are not only conduits for the passive transport of inflammatory cells by lymph flow, but also via chemoattractants have the capacity to actively recruit them and influence their migration. While this may contribute to the shaping of the immune response, inflammatory cells are also rich sources of lymphangiogenic growth factors and cytokines, which effects on the growth and structure of the lymphatic system.

The pro-inflammatory function of lymphatic vessels reaches clinical relevance when avascular tissue becomes vascularized. A particularly relevant setting is the lymphangiogenic vascularization of cornea transplants. The role of lymphatic vessels and lymphangiogenesis in tissue and organ rejection is reviewed in Chap. 10.

The colinearity of blood vessels, lymphatic vessels, and peripheral nerves suggests common patterning principles may guide and control the growth and bifurcation of these tissues. Indeed, neuronal axon guidance molecules have been demonstrated to control blood and now also lymphatic vessel development. This fascinating aspect of tissue morphogenesis is highlighted in Chap. 11.

Lymphatic vessels are an organ present in all higher vertebrates. In particular, the zebrafish has proven to be a most valuable model organism, because it is amenable to large-scale forward genetic screens, and the rapid development of the transparent embryo allows convenient life imaging. As depicted in Chap. 12, these features have been used to identify novel gene functions during lymphangiogenesis and have also allowed a detailed description of the cellular events that lead to the formation of a fully patent lymphatic vascular system in the early zebrafish embryo.

Clearly, in mammals imaging of either lymphatic vessel development or vessel function in the adult is immeasurably more complicated. Nevertheless, an in-depth understanding of lymph vessel biology relies on the spatial interrogation of the entire developing vasculature as well as the availability of live imaging. Chapter 13 outlines the application of light sheet microscopy for the analysis of embryonic vessels and discusses different imaging modalities, in particular two-photon microscopy in conjunction with fluorescent proteins and optical chambers for live imaging in the mouse.

Understanding the physiology and pathophysiology of lymphatic vessels at various levels ranging from a clinical perspective to the cellular and even molecular level has been an important goal, for both clinicians but also biomedical researchers not in contact with patients. Particularly in recent years, there is more frequent cross talk between clinicians and lab researchers, leading to an increasingly better understanding of the often complex features of lymphatic syndromes. In the spirit of this interdisciplinary comprehensive approach, the concluding review in Chap. 14 is focussed on primary lymphedema.

References

Oliver, G., & Detmar, M. (2002). The rediscovery of the lymphatic system: Old and new insights into the development and biological function of the lymphatic vasculature. *Genes and Development, 16*, 773–783.

Sabin, F. R. (1902). On the origin of the lymphatic system from the veins and the development of the lymph hearts and thoracic duct in the pig. *The American Journal of Anatomy, 1*, 367–389.

Sabin, F. R. (1916). The method of growth of the lymphatic system. *Science, 44*, 145–158.

Srinivasan, R. S., Dillard, M. E., Lagutin, O. V., Lin, F. J., Tsai, S., Tsai, M. J., et al. (2007). Lineage tracing demonstrates the venous origin of the mammalian lymphatic vasculature. *Genes and Development, 21*, 2422–2432.

Chapter 2
Transcriptional Control of Lymphatic Endothelial Cell Type Specification

Ying Yang and Guillermo Oliver

Abstract The lymphatic vasculature is the "sewer system" of our body as it plays an important role in transporting tissue fluids and extravasated plasma proteins back to the blood circulation and absorbs lipids from the intestinal tract. Malfunction of the lymphatic vasculature can result in lymphedema and obesity. The lymphatic system is also important for the immune response and is one of the main routes for the spreading of metastatic tumor cells. The development of the mammalian lymphatic vasculature is a stepwise process that requires the specification of lymphatic endothelial cell (LEC) progenitors in the embryonic veins, and the subsequent budding of those LEC progenitors from the embryonic veins to give rise to the primitive lymph sacs from which the entire lymphatic vasculature will eventually be derived. This process was first proposed by Florence Sabin over a century ago and was recently confirmed by several studies using lineage tracing and gene manipulation. Over the last decade, significant advances have been made in understanding the transcriptional control of lymphatic endothelial cell type differentiation. Here we summarize our current knowledge about the key transcription factors that are necessary to regulate several aspects of lymphatic endothelial specification and differentiation.

2.1 Overview of the Stepwise Process Leading to the Formation of the Lymphatic Network

Mammals have two interdependent circulatory systems—the blood vasculature and the lymphatic vasculature. Although detailed descriptions of the blood vascular system were available as early as the sixth century BC, those of the lymphatic vasculature were in the seventeenth century AD by Asellius. In contrast to the

Y. Yang • G. Oliver (✉)
Department of Genetics, St. Jude Children's Research Hospital, Memphis, TN, USA
e-mail: Guillermo.oliver@stjude.org

F. Kiefer and S. Schulte-Merker (eds.), *Developmental Aspects of the Lymphatic Vascular System*, Advances in Anatomy, Embryology and Cell Biology 214,
DOI 10.1007/978-3-7091-1646-3_2,

function of the blood vasculature in transporting blood throughout the body, the lymphatic vasculature is essential for maintaining interstitial fluid homeostasis. The main physiological functions of the lymphatic vasculature include draining and returning fluid from the extracellular tissue spaces back to the blood circulation, absorbing lipids from the intestinal tract and tissues, and transporting immune cells to lymphoid organs.

During the formation of the blood vascular network, the Notch signaling pathway is required to promote arterial cell fate differentiation (De Val and Black 2009; Kokubo et al. 2005; Lawson et al. 2001). On the other hand, the orphan nuclear receptor *COUP-TFII* promotes venous fate differentiation by inhibiting Notch signaling and other arterial specification genes (You et al. 2005).

Studies on the lymphatic network in the past decade have brought important advances in the understanding of the molecular events that lead to the development of the lymphatic vasculature. At least in the mammalian embryo, the formation of the lymphatic vasculature is a stepwise process that is closely associated with the venous vasculature. The pioneering work by Florence Sabin at the beginning of the last century proposed that the lymphatic vasculature arises from the embryonic veins (Sabin 1902). Detailed lineage tracing analysis performed in mouse embryos almost a century later confirmed Sabin's prediction that lymphatic endothelial cells (LECs) have a venous origin (Srinivasan et al. 2007). Therefore, a key prerequisite for the genesis of the lymphatic network is the prior formation of the blood vasculature.

In the mouse embryo, the process leading to the formation of the entire lymphatic vascular network starts inside the cardinal vein (CV) and intersomitic vessels at around embryonic day 9.5 (E9.5) (Srinivasan et al. 2007; Wigle and Oliver 1999; Yang et al. 2012; Hagerling et al. 2013) (Fig. 2.1a). As discussed in detail later, at this stage a subpopulation of venous ECs become specified into the lymphatic lineage by acquiring a very specific molecular footprint. Concomitant with their stepwise loss of venous fate, these ECs will gain lymphatic fate and as such they should be considered LEC progenitors. The formation of the entire lymphatic vascular network is mediated by intermediate structures called lymph sacs. Lymph sacs are formed when most of the LEC progenitors bud off from the veins and migrate into the surrounding mesenchyme. Electron microscopy and immunostaining studies have revealed that adhesion junctions between the venous endothelial cells (VECs) and LEC progenitors are important for maintaining vein integrity during the budding process (Yang et al. 2012; Yao et al. 2012) (Fig. 2.2). As part of the stepwise differentiation process, as they bud off, LEC progenitors start expressing additional genes (e.g., podoplanin) and differentiate into more mature LECs outside the veins (Yang et al. 2012) (Fig. 2.1a, b). Importantly, at the level of the junction of the jugular and subclavian veins, a small fraction of LEC progenitors will remain inside the CV and contribute to the formation of the lymphovenous valves, which are the main valves responsible for the unidirectional return of lymph fluid into the blood circulation (Srinivasan and Oliver 2011) (Fig. 2.1b, d).

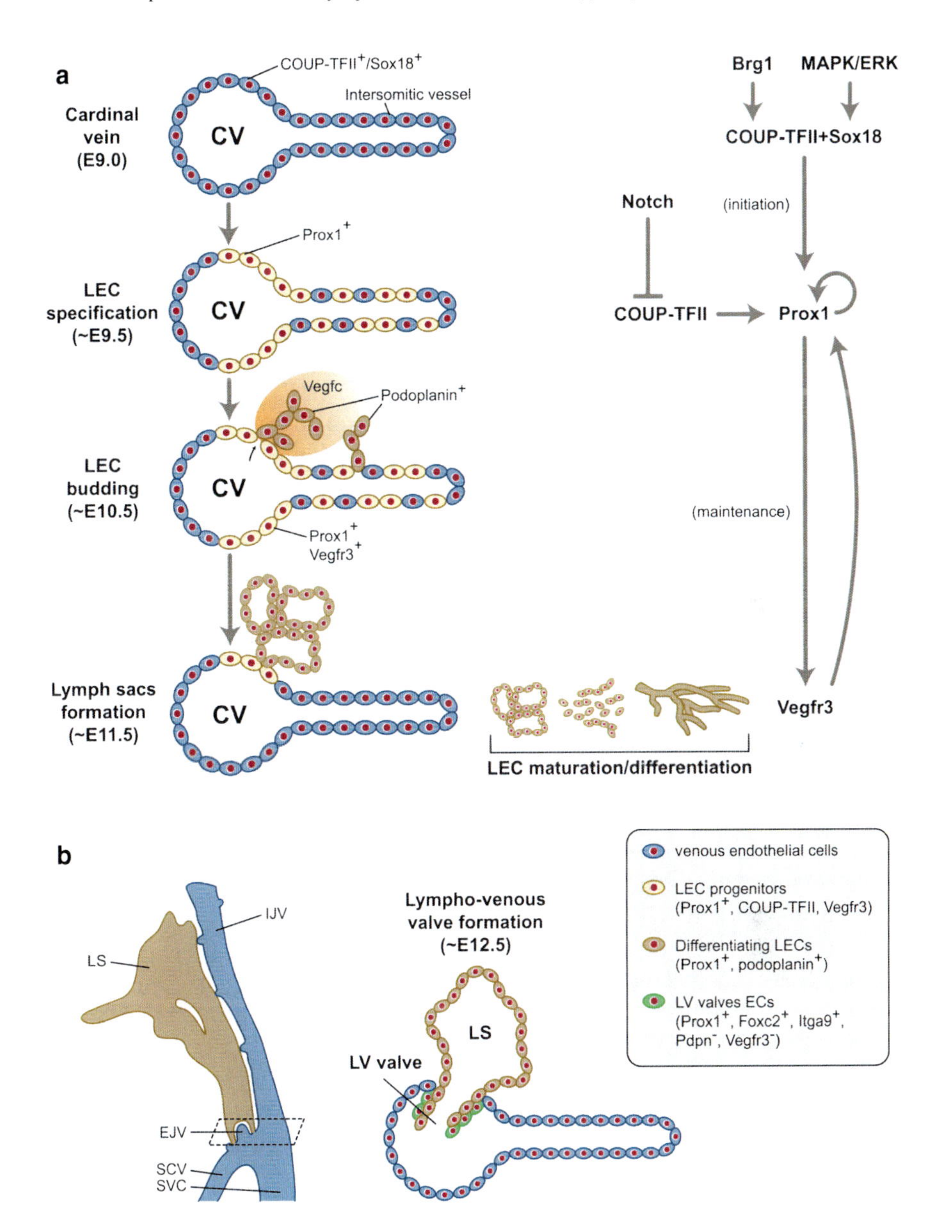

Fig. 2.1 Schematic representation of the stepwise process leading to the formation of the mammalian lymphatic vasculature. (**a**) The cardinal vein (CV) is the source from where LECs are going to be specified. Early during development, the CV expresses the transcription factors *COUP-TFII* and *Sox18*. Together, the activity of these factors will be necessary to induce the expression of *Prox1* in a subpopulation of venous ECs. The initiation of *Prox1* expression at around E9.5 is an indication that LEC specification started and the venous *Prox1*-expressing ECs should be considered as LEC progenitors. Approximately a day later (E10.5), most of those progenitors will start to bud off from the CV. This process is guided by the graded expression of *Vegfc* in the surrounding mesenchyme. LECs will bud off from the CV and intersomitic vessels as a chain of interconnected cells. *Vegfr3* expression in LECs is maintained by *Prox1*. As soon as LECs bud from the vein they start to express podoplanin, an indication that lymphatic maturation

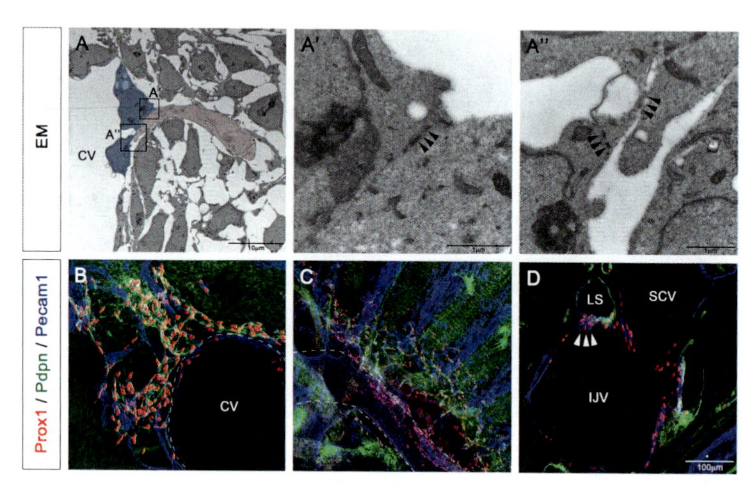

Fig. 2.2 *Prox1*-expressing LEC progenitors bud from the CV (**a**, **a**′, **a**″). EM analysis of E10.5 *Prox1*$^{+/LacZ}$ embryos showing that *Prox1*-expressing cells (*pink*) exit the CV via an active budding mechanism. The budding cell is attached to the venous endothelial cells (*blue*) by adhesion junctions (*arrowheads* in (**a**′) and (**a**″)). Panels (**a**′, **a**″) are high-power magnifications of the black boxed area in (**a**). (**b**) Podoplanin expression is only detected once *Prox1*-expressing LEC progenitors fully exit the CV (*dashed line*) at around E11.5. (**c**) *Prox1*-expressing LEC progenitors that bud off from the CV (*dashed line*) and ISVs migrate as an interconnected group of cells dorsally and longitudinally into the surrounding mesenchymal tissue at approximately E11.0 in the anterior region of the embryo. (**d**) A small subpopulation of *Prox1*-expressing venous ECs remain in the veins and form the lymphovenous valves at the junction of the jugular and subclavian veins at E15.5 (*arrowheads* in (**d**)). These valve cells are negative for podoplanin. *CV* cardinal vein, *IJV* internal jugular vein, *SCV* subclavian vein. *Scale bars*: 10 μm (**a**), 1 μm (**a**′, **a**″), 100 μm (**b**–**d**)

Vascular endothelial growth factor C (*Vegfc*) signaling is indispensable for the budding process. A newly identified player in this pathway, *Ccbe1*, is also required for LEC budding (Bos et al. 2011; Hogan et al. 2009). In the absence of either molecule, LEC progenitors fail to bud off from the embryonic veins (Karkkainen et al. 2004; Hagerling et al. 2013; Bos et al. 2011). Recent studies have shown that as LECs bud off from the veins, they remain interconnected to each other during migration (Yang et al. 2012; Hagerling et al. 2013) (Fig. 2.2c). These interconnected LECs form strings that gradually organize into the lumenized structures called lymph sacs. Earlier studies argued that by sprouting and remodeling, the lymph sacs and the early lymphatic plexus give rise to the mature

---

Fig. 2.1 (continued) and differentiation is progressing. As LECs bud, they will start to form the different lymph sacs (starts at around E11.5), intermediate structures from which upon maturation and differentiation, the entire lymphatic network will be formed. (**b**) Although most *Prox1*-expressing LEC progenitors will move out from the veins, a small fraction located at the junction of the internal (IJV) or external jugular veins (EJV) with the subclavian vein (SCV) will remain and upon their intercalation with lymph sac (LS) LECs, they will help to form the lymphovenous valves

lymphatic vasculature, including both superficial lymphatics and the thoracic duct (TD) (Sabin 1902; van der Putte 1975; Oliver 2004). However, digital three-dimensional reconstructions using ultramicroscopy suggest a different model. According to this new model, simultaneous with the formation of primitive lymph sacs, LECs accumulate at the first lateral branch of the intersomitic vessels to form a peripheral longitudinal lymphatic vessel that will give rise to the superficial lymphatic plexus while the primitive lymph sacs will later develop into the TD (Hagerling et al. 2013). After the primitive lymphatic vessels are formed, they will further differentiate into larger collecting vessels and smaller capillaries, which are the two predominant types of vessels of the lymphatic vasculature. Lymphatic capillaries absorb interstitial fluid and carry this lymph toward the larger collecting lymphatic vessels. In the collecting lymphatics, valves form to prevent the backflow of the lymph and separate the lymphatic vessels into functional units called lymphangions (Sacchi et al. 1997; Casley-Smith 1980).

2.2 Transcriptional Control of Lymphatic Specification and Differentiation

2.2.1 Sox18 *and* COUP-TFII *Initiate Lymphatic Endothelial Cell Transcriptional Profiling*

So far, only a few transcription factors critical during LEC specification and differentiation have been identified, and how these genes are regulated and how they interact with other critical signaling pathways (e.g., *Vegfc–Vegfr3*, MAPK/ERK, and Notch signaling) remain to be fully elucidated.

Many transcription factors play an important role in regulating blood endothelial cell (BEC) fate differentiation. Given the common origin and close relationship of venous and lymphatic endothelial cells, it is not surprising that at least some venous endothelial transcription factors are also required for the development of the lymphatic vasculature. For instance, *Sox18* encodes an SRY-type HMG box transcription factor and is a member of the SOX gene family. Mutations in *SOX18* in humans are associated with hypotrichosis–lymphedema–telangiectasia (Irrthum et al. 2003). *Sox18* is expressed in the vascular endothelium and hair follicles in mouse embryos (Pennisi et al. 2000b). Point mutations in *Sox18* result in cardiovascular and hair follicle defects in mice; as a consequence, these mice are known as *ragged* mice (Carter and Phillips 1954; Slee 1957a, b). On the other hand, *Sox18* knockout mice display a milder coat defect and no obvious cardiovascular defects in a mixed genetic background (Pennisi et al. 2000a).

Some evidence supports that other Sox transcription factors have redundant functions. For example, *Sox18* and *Sox7* can compensate for the loss of each other in arteriovenous specification in zebrafish (Herpers et al. 2008; Cermenati et al. 2008; Pendeville et al. 2008). In the mouse, the expression of *Sox18* in

Table 2.1 Key players in the lymphatic endothelial transcriptional network

Gene	Binding sites in enhancers or promoters	Binding site validation	Interaction protein	References
Sox18	Ets1; Egr1	N/A	N/A	Deng et al. (2013)
COUP-TFII	Brg1	ChIP	Prox1	Davis et al. (2013), Lee et al. (2009), Yamazaki et al. (2009)
Prox1	SOXF (Sox18); COUP-TFII; miR-181; miR-31	EMSA; ChIP; Tg; TA	COUP-TFII; Ets2; NF-κB	Francois et al. (2008), Srinivasan et al. (2010)
Vegfr3	Prox1; Est2; Tbx1	ChIP; TA	N/A	Yoshimatsu et al. (2011), Chen et al. (2010)
Nrp2	Sp1 (COUP-TFII)	ChIP; TA	N/A	Lin et al. (2010)
Foxc2	N/A	N/A	NFATc1	Norrmen et al. (2009)
NFATc1	N/A	N/A	Foxc2	Norrmen et al. (2009)
Gata2	ETS; E box (Tal1)	EMSA; Tg	N/A	Khandekar et al. (2007)

The list includes validated binding sites and interaction proteins for lymphatic endothelial transcription factors. Relevant references for each promoter/enhancer and interaction protein are listed *EMSA* gel shift assay, *ChIP* chromatin immunoprecipitation, *Tg* transgenic embryos, *TA* trans-activation assays in cell culture, *N/A* not applicable

vascular endothelial cells starts as early as E7.5 in the allantois and yolk-sac blood islands (Pennisi et al. 2000b). *Sox18* expression in ECs in the CV is detected at around E9.0, but gets downregulated in LECs at around E14.5 and is not maintained in the adult lymphatic vasculature (Francois et al. 2008). *Sox18* is indispensible for generating venous LEC progenitors via its activation of the critical transcription factor *Prox1* (prospero homeobox 1, *Drosophila prospero*-related vertebrate gene) in VECs (Francois et al. 2008). However, this lack of LEC progenitors in *Sox18* mutant mice has only been observed in a pure C57BL/6 (B6) background; *Sox7* and *Sox17* functionally substitute for *Sox18* in a mixed background (Hosking et al. 2009). A 4 kb *Prox1* promoter region that is sufficient to recapitulate *Prox1* expression in vivo contains two conserved *SoxF* binding sites that in vitro can be bound directly by *Sox18* (Francois et al. 2008) (Table 2.1). These results show that *Sox18* is a direct upstream activator of *Prox1* in ECs in the CV, and as such, also an important early player in the acquisition of LEC progenitor fate. Previously, the MAPK/ERK pathway was shown to regulate lymphatic vessel growth by modulating *Vegfr3* expression in mice (Ichise et al. 2010). A recent report revealed that MAPK/ERK signaling might be responsible for activating *Sox18* expression in the CV (Deng et al. 2013). Normally, induction of *Prox1* expression by *Sox18* occurs only in the embryonic veins (i.e., not in the arteries). This specificity could be because arteries fail to express certain specific factors required to activate *Prox1* in combination with *Sox18* (Francois et al. 2008) or because they express certain genes (repressors) that inhibit *Prox1* activation. The finding that gain of function of the MAPK/ERK signaling component RAF1 in ECs induces an abnormal expression level of *Sox18* in the veins and dorsal aorta, which in turn activates *Prox1* expression in these vessels (Deng et al. 2013), argues that aberrantly activated MAPK/ERK signaling either inhibits arterial-specific repressor(s) or turns on the expression of venous *Sox18* coactivators in the arteries.

As mentioned above, in addition to *Sox18*, other coactivators are necessary to induce *Prox1* expression in the veins and initiate the specification of LEC progenitors. The chicken ovalbumin upstream promoter-transcription factor II (*COUP-TFII*) is a venous cofactor that can bind to the *Prox1* promoter and activate its expression (Srinivasan et al. 2010). *COUP-TFII* is a member of the steroid/thyroid hormone receptor superfamily and is highly expressed in the mesenchymal tissue as well as in the blood vascular endothelium during development (Pereira et al. 1995). Within the vasculature, its expression is restricted to the veins, where its activity is required to promote and maintain venous identity by inhibiting Notch activity in VECs, thus blocking the arterial transcriptional program (You et al. 2005). A recent study reported that *COUP-TFII* activity in the veins can be regulated epigenetically by the chromatin-remodeling enzyme gene *Brg1*, a member of the SWI/SNF protein family (Davis et al. 2013). *Brg1* remodels the chromatin structure of the *COUP-TFII* promoter region by direct binding, thereby preventing the access of the transcriptional machinery to that region (Davis et al. 2013). Interestingly, *COUP-TFII* is not only essential for venous cell fate differentiation but is also involved in the specification of LEC progenitors during lymphatic vasculature development. Deletion of *COUP-TFII* expression from LEC progenitors causes a drastic reduction in the number of LECs (Srinivasan et al. 2007). Moreover, *COUP-TFII* directly binds conserved binding sites located approximately 9.5 kb upstream of the *Prox1* open reading frame (ORF) (Srinivasan et al. 2010) (Table 2.1). These results argue that *COUP-TFII* is another direct in vivo regulator of *Prox1* expression during the early phases of LEC specification in the CV (Fig. 2.1). The maintenance of *Prox1* expression in LECs also requires *COUP-TFII* (see below). In addition, conditional inactivation of *COUP-TFII* at different embryonic stages revealed that its activity is not only essential for the specification of LEC progenitors but also for the sprouting of dermal lymphatic capillaries prenatally (Lin et al. 2010). In the absence of *COUP-TFII*, the lymphatic capillaries failed to form filopodial extensions projecting from the vessels. This function of *COUP-TFII* is via direct transcriptional regulation of *Nrp2*, a coreceptor for *Vegfc* in LECs (Lin et al. 2010; Xu et al. 2010) (Table 2.1). However, although *COUP-TFII* is expressed in LECs throughout life, its activity is not required to maintain quiescent lymphatic vessels in the adult, as normal lymphatic function remains intact when *COUP-TFII* is inactivated in adult mice (Lin et al. 2010). Thus, *Sox18* and *COUP-TFII* are two transcription factors crucial to initiate *Prox1*-mediated LEC progenitor specification in the veins. Although both *Sox18* and *COUP-TFII* bind to the *Prox1* promoter to induce its expression, either one is sufficient to activate *Prox1* expression by itself. However, how these two transcription factors interact with each other in this process remains uncertain.

2.2.2 Prox1 *Regulates LEC Progenitor Specification and Differentiation*

It is widely accepted that LEC specification begins with *Prox1* expression in VECs at around E9.5 (Oliver 2004) (Fig. 2.2). *Prox1* (prospero homeobox 1, vertebrate gene related to *Drosophila prospero*) (Oliver et al. 1993) was the first lymphatic endothelial transcription factor to be identified (Wigle and Oliver 1999). Although *Prox1* is expressed in many other tissue types such as the central nervous system, lens, heart, liver, and pancreas (Oliver et al. 1993), its expression in the vascular system is restricted to lymphatic ECs. *Prox1* begins to be expressed in a subpopulation of ECs on the anterior CV (LEC progenitors) at around E9.5 and its expression is maintained in LECs through life (Wigle et al. 2002; Wigle and Oliver 1999). Functional inactivation of *Prox1* in mice showed that in the absence of *Prox1* expression, the embryo was devoid of the entire lymphatic vasculature and died at around E14.5 (Wigle and Oliver 1999). It was later shown that conditional deletion of *Prox1* at any developmental or postnatal stage leads to the loss of lymphatic-specific gene expression and the concomitant upregulation of BEC-specific genes in LECs (Johnson et al. 2008). These findings show that *Prox1* activity is required for the formation of LEC progenitors and for the maintenance of LEC identity at all developmental stages, including adulthood. Furthermore, it was recently shown that *Prox1* activity is also necessary for LECs to bud from the embryonic veins (Yang et al. 2012). Finally, several in vitro gain-of-function studies indicated that ectopic expression of *Prox1* in BECs is able to initiate a mature lymphatic-specific gene expression (e.g., *Vegfr3* and *podoplanin*) while suppressing BEC-specific genes (Hong et al. 2002; Petrova et al. 2002). Taken together, these studies indicate that *Prox1* is a key transcriptional regulator not only during the specification of LEC progenitors but also during LEC differentiation.

As *Prox1* is a central regulator of lymphatic endothelial transcription, its expression is tightly regulated. Several lines of evidence suggest that *Prox1* dosage is crucial for its function. In most genetic backgrounds, haploinsufficiency of *Prox1* causes perinatal death and pups exhibit characteristics of lymphatic dysfunction (e.g., chylothorax and chylous ascites) (Harvey et al. 2005). In other backgrounds, a small proportion of *Prox1* heterozygous animals survive to adulthood; however, their lymphatic vessels are mispatterned and leaky, and mice develop adult-onset obesity (Harvey et al. 2005). All *Prox1* heterozygote embryos display edema and occasionally develop blood-filled lymphatics (Srinivasan and Oliver 2011; Harvey et al. 2005). In these embryos, the number of LEC progenitors is significantly decreased and lymphovenous valves are absent (Srinivasan and Oliver 2011). Detailed analyses of *Prox1* null embryos in which the ORF of *Prox1* was replaced with either *LacZ* or *GFP* reporter gene constructs revealed that *Prox1* is essential for maintaining its own expression in LEC progenitors; this autoregulation is crucial for LEC identity (Wigle et al. 2002; Srinivasan et al. 2010). At the molecular level, in vitro studies suggest a physical interaction between *COUP-TFII* and *Prox1* in LECs (Lee et al. 2009; Yamazaki et al. 2009). Similarly, a recent study demonstrated that *COUP-TFII* homodimers induce VEC fate by repressing the Notch target genes *HEY1/2*, whereas COUP-TFII/Prox1 heterodimers induce or are permissive for the expression of a subgroup of LEC-specific genes (Aranguren et al. 2013). Comprehensive phenotype analyses showed that in *COUP-TFII/Prox1* double heterozygotes embryos, the loss of *COUP-TFII* aggravates the lymphatic defects of *Prox1* heterozygotes (Srinivasan et al. 2010). Compared with *Prox1* heterozygote embryos, the number of LECs and the expression level of *Prox1* are further reduced in *COUP-TFII/Prox1* double heterozygotes (Srinivasan et al. 2010). This result argues that the activity of *COUP-TFII* is required to maintain *Prox1* expression in LECs and that the amount of COUP-TFII/Prox1 protein complex is important to regulate *Prox1* expression in a dosage-dependent manner (Srinivasan et al. 2010; Srinivasan and Oliver 2011). In addition, embryos with a mutated *Prox1* nuclear hormone receptor-binding site in which the interaction between *Prox1* and *COUP-TFII* is abolished displayed similar LEC specification defects (Srinivasan et al. 2010). Taken together, these results support that the COUP-TFII/Prox1 interaction is required to maintain *Prox1* expression in LEC progenitors and, therefore, LEC identity during the LEC specification stage. Once LEC progenitors are specified and start to differentiate as they bud off from the CV, *COUP-TFII* activity is no longer required to maintain *Prox1* expression (Srinivasan et al. 2010).

Since *COUP-TFII* is a crucial regulator in both venous and lymphatic specification (by suppressing Notch activity and triggering *Prox1* expression in the veins), it can be speculated that Notch signaling may also be involved in lymphatic vasculature development. It is well known that Notch signaling is critical for the development of the blood vasculature, including arteriovenous specification and angiogenic sprouting (Gridley 2010; Roca and Adams 2007). However, until recently the role of Notch signaling in lymphatic vasculature development remained a matter of debate, mainly because there were no conclusive data demonstrating the presence of either in vivo expression of Notch pathway components in LECs or in vivo functional studies. For example, in vivo deletion of *Rbpj* (a key mediator of Notch signaling) in ECs did not result in LEC specification defect in the embryos (Srinivasan et al. 2010). On the other hand, in vitro studies showed that ectopic expression of an activated Notch receptor in cultured LECs repressed *Prox1*, *COUP-TFII*, and the mature lymphatic marker podoplanin through Hey proteins (downstream effectors of Notch signaling) (Kang et al. 2010). Likewise, addition of soluble Jag1 or Dll4 recombinant protein (Notch ligands) into the culture medium suppressed the expression of *Prox1*, *COUP-TFII*, and podoplanin also through Hey proteins (Kang et al. 2010). These data suggest that in vitro Notch signaling inhibits LEC fate. Another study also proposed that blocking Notch promotes LEC sprouting in vitro. This Notch inhibition-induced lymphangiogenesis required *Vegfr2* and *Vegfr3* signaling (Zheng et al. 2011). Thus, these data indicate that in vitro Notch signaling is involved in both the specification and sprouting of LECs. Contradictory to these results, another study demonstrated that by treating neonatal mouse tail dermis, ears, and retinas with blocking antibodies targeting Notch1 and Dll4, lymphatic vessel sprouting and growth were impaired (Niessen et al. 2011). Recently, some in vivo studies have supported that Notch activity plays a role during embryonic lymphatic development (Murtomaki et al. 2013). These studies showed that Notch signaling components are present in LECs during embryonic development. Also, removal of Notch1 or disturbing Notch transcription in LECs leads to an increase in the number of LEC progenitors and to larger lymph sacs (Murtomaki et al. 2013). These results argue that Notch signaling acts as a negative regulator of LEC specification by repressing *Prox1* expression. Taken together, these findings indicate that Notch signaling is a negative regulator of LEC fate decisions during lymphatic vasculature development (Fig. 2.1).

Because of the limited number of LEC progenitors on the embryonic veins and the lack of specific surface markers to sort these cells, the identification of direct in vivo target genes of *Prox1* in these cells has been challenging. Several lymphatic genes are regulated by *Prox1* expression in vitro (Mishima et al. 2007; Hong et al. 2002; Fritz-Six et al. 2008; Sabine et al. 2012; Harada et al. 2009; Shin et al. 2006). Some studies have suggested that the main receptor of *Vegfc* signaling, *Vegfr3*, is a downstream target of *Prox1*. *Vegfr3* is expressed in BECs and is essential for blood vasculature development. *Vegfr3* null mouse embryos die at around E10.0 with severe defects in remodeling of the primary vessel networks (Dumont et al. 1998). In wild-type embryos LECs start to express *Vegfr3* after E10.5. *Vegfc* is the most well-characterized *Vegfr3* ligand, and in the absence of *Vegfc* signaling, *Prox1*$^+$ LEC progenitors fail to bud off from the embryonic veins (Karkkainen et al. 2004). Previous microarray data indicate that ectopic expression of *Prox1* in cultured VECs leads to a significant increase in *Vegfr3* expression (Hong et al. 2002; Petrova et al. 2002). Furthermore, during inflammation-induced lymphangiogenesis, *Prox1* transcriptionally regulates *Vegfr3* expression by binding to its promoter together with NF-κB or Ets2 in vitro (Flister et al. 2009; Yoshimatsu et al. 2011). These in vitro data suggest that *Vegfr3* is a direct target of *Prox1* and that other coactivators such as *COUP-TFII*, NF-κB, or Ets2 may be involved in this regulatory process (Table 2.1). Our own unpublished data recently confirmed that *Vegfr3* is a direct in vivo target of *Prox1* in a dosage-dependent manner (Table 2.1). We determined that Prox1 maintains Vegfr3 expression in LEC progenitors and the number of LEC progenitors. Furthermore, the expression level of Prox1 in those cells is further reduced in Vegfr3+/−; Prox1+/− embryos, revealing the existence of a regulatory feedback loop between Prox1 and Vegfr3. Therefore, in addition to *COUP-TFII*, *Vegfr3* also regulates *Prox1* expression during the early specification and differentiation of LEC progenitors.

Besides transcriptional regulation, at least in vitro *Prox1* expression is also controlled by posttranscriptional regulation. It has been reported that lysine 556 is the major sumoylation site for *Prox1* and that sumoylation of *Prox1* influences its activity (Pan et al. 2009; Shan et al. 2008). In addition, in vitro data suggest that microRNAs regulate *Prox1* levels in LECs, as *Prox1* expression is negatively regulated by *miR-181* or *miR-31* in cultured LECs (Kazenwadel et al. 2010; Pedrioli et al. 2010) (Table 2.1). However, the in vivo function of the posttranscriptional regulation of *Prox1* remains unknown. Interestingly, blood flow plays a significant role in modulating lymphatic identity in vivo. In this context, the expression of *Prox1* is rapidly lost when lymphatic vessels are exposed to high shear rates from blood flow, leading to the loss of lymphatic identity (Chen et al. 2012). Taken together, these results highlight that as a central player during LEC specification and differentiation, the level of *Prox1* is strictly regulated by numerous environmental and genetic factors. More transcription factors and signaling pathways that affect *Prox1* expression remain to be discovered.

2.3 Not All *Prox1*-Expressing LEC Progenitors Will Leave the CV

As discussed above, *Prox1* activity is required for the specification of LEC progenitors and for those progenitors to bud off from the CV. However, a recent study has identified a small subpopulation of *Prox1*-expressing LEC progenitors that will remain in the veins and help to form the lymphovenous valves (Srinivasan and Oliver 2011) (Fig. 2.1b, d). These cells are located at the junction of the jugular and subclavian veins and will not acquire LEC features (e.g., will not express podoplanin). Instead, they express an additional set of markers such as *Foxc2* and *Itga9* (Fig. 2.1b). Following intercalation with a subpopulation of venous ECs they will form the lymphovenous valves (Figs. 2.1 and 2.2). The formation of *Prox1*-expressing venous ECs and the derived lymphovenous valves is also dependent on *Prox1* activity, as these valves are absent in *Prox1* heterozygous mice (Srinivasan and Oliver 2011). This defect is a consequence of defective maintenance of *Prox1* expression in LEC progenitors, which is promoted by a reduction in the formation of the COUP-TFII/Prox1 complex (Srinivasan and Oliver 2011). Together, these results support that *Prox1*-expressing venous ECs are the source of cells that will produce both LECs progenitors and lymphovenous valves. However, what makes some *Prox1*-expressing ECs remain on the vein remains to be determined.

2.4 *Foxc2* Is an Essential Regulator of Lymphatic Maturation and Valve Formation

Once the specified, mature LECs form the primitive lymph sacs and lymphatic plexus, they will differentiate further and give rise to the collecting lymphatic vessels and lymphatic capillaries. The formation of the lymphatic valves is an important step during the maturation of the primitive lymphatic plexus into collecting lymphatics. *Foxc2*, a member of the forkhead/winged-helix family of transcription factors, is one of the main players in the regulation of this critical step. In humans, point mutations in *FOXC2* have been identified as the cause of lymphedema-distichiasis (LD) (Fang et al. 2000; Brice et al. 2002). A similar phenotype was observed in *Foxc2*$^{+/-}$ mice (additional row of eyelashes, increased number of lymph nodes, and lymph backflow), suggesting that *Foxc2*$^{+/-}$ is a suitable mouse model for LD (Kriederman et al. 2003). *Foxc2* is necessary for lymphatic patterning, lymphatic valve formation, and mural cell recruitment during the maturation stage (Petrova et al. 2004). Inactivation of *Foxc2* results in dilated lymphatic capillaries that become ectopically covered with smooth muscle actin-positive perivascular cells, whereas normal lymphatic capillaries lack mural cell coverage. It has been suggested that *Foxc2* and *Vegfr3* cooperate during the patterning of the lymphatic vasculature, and *Foxc2* presumably functions downstream of *Vegfr3* (Petrova et al. 2004). Although *Foxc2* is normally expressed in LECs from E9.5 to adult stages, its activity is not required for the budding and migration of LEC progenitors from the embryonic veins or the formation of lymph sacs (Dagenais et al. 2004). Studies have shown that *Foxc2* is also essential for the maturation of collecting lymphatics (Norrmen et al., 2009). Collecting lymphatics start to form around E14.5–15.5, and during their maturation markers for lymphatic capillaries such as *Prox1*, *Vegfr3*, and *Lyve1* get downregulated and valves start to form. In the absence of *Foxc2*, the expression of these markers remains high and valves do not form; as a consequence, the primary lymphatic plexus fails to mature into functional collecting lymphatics (Norrmen et al. 2009). Coimmunoprecipitation assays and genome-wide location mapping revealed that *Foxc2* physically interacts with *NFATc1*, a regulator of cardiac valve development (Chang et al. 2004; de la Pompa et al. 1998; Ranger et al. 1998), and functionally cooperates with calcineurin/*NFATc1* signaling in transcriptional regulation during the development of collecting lymphatics (Norrmen et al. 2009) (Table 2.1). Importantly, calcineurin/*NFATc1* signaling is required for normal lymphatic vascular patterning and LEC-specific gene expression during development. Blocking calcineurin/*NFATc1* signaling with the calcineurin inhibitor cyclosporine A in utero results in the loss of podoplanin and Fgfr3 expression in LECs (Kulkarni et al. 2009). Furthermore, *Foxc2*-calcineurin/*NFATc1* signaling is not only important during collecting lymphatic vessel maturation but also indispensible for the formation and maintenance of lymphatic valves. The formation of lymphatic valves starts around E16.0, which is indicated by elevated *Prox1* and *Foxc2* expression in lymphatic valve-forming cells (Sabine et al. 2012). *Foxc2*-calcineurin/*NFATc1* signaling is activated in developing lymphatic valves, as indicated by the accumulation of nuclear *NFATc1* in lymphatic valve-forming cells. Retrograde lymph flow is observed in *Foxc2*$^{-/-}$ embryos because of the complete absence of lymphatic valves (Petrova et al. 2004). Removal of calcineurin in ECs is also sufficient to affect the formation of a lymphatic valve territory (Sabine et al. 2012). In addition, the inactivation of calcineurin at any developmental stage results in lymphatic valve defects, indicating that *Foxc2*–calcineurin/*NFATc1* signaling is not only crucial for the initiation of valve formation but also required for the maintenance of lymphatic valves (Sabine et al. 2012). Besides calcineurin/*NFATc1* signaling, the gap junction protein *Cx37* is also essential for the assembly of the lymphatic valve territory. The clusters of lymphatic valve-forming cells were absent in *Cx37*-

knockout mice, resulting in the absence of lymphatic valves (Kanady et al. 2011; Sabine et al. 2012). In vitro flow analyses revealed that the expression of *Cx37* and calcineurin/*NFATc1* activation in LECs is regulated by oscillatory fluid shear stress in a *Prox1*- and *Foxc2*-dependent manner and that *Cx37* depletion significantly decreases calcineurin/*NFATc1* activation (Sabine et al. 2012). In vivo, *Cx37* was almost completely absent in LECs of *Foxc2*$^{-/-}$ embryos (Sabine et al. 2012). Taken together, these results support that *Prox1*, *Foxc2*, and shear stress coordinate the expression of *Cx37*, which in turn activates calcineurin/*NFATc1* signaling in the lymphatic valve-forming cells during lymphatic valve morphogenesis.

2.5 Additional Transcription Factors Involved in Lymphatic Development

Tbx1, a member of a conserved family of transcription factors that share a common T-box DNA-binding domain, has been recently identified as a gene whose activity is necessary during lymphatic development (Chen et al. 2010). *Tbx1* is associated with the DiGeorge syndrome; however, lymphatic defects are rarely reported in patients with this syndrome (Yagi et al. 2003; Mansir et al. 1999). During mouse development, deletion of *Tbx1* from ECs leads to embryonic edema and postnatal lethality between 2 and 4 days after birth because of abdominal chylous ascites (Chen et al. 2010). These mice have severely reduced lymphatic vessel density in the heart, diaphragm, and skin and lack the entire gastrointestinal lymphatic vasculature (Chen et al. 2010). Conditional inactivation of *Tbx1* at different developmental stages revealed that *Tbx1* activity is required until E14.5 for the formation of the mesenteric lymphatic vasculature (Chen et al. 2010). Mechanistically, chromatin immunoprecipitation analysis has shown that *Tbx1* binds to conserved T-box-binding elements in the *Vegfr3* promoter to activate its expression (Chen et al. 2010) (Table 2.1).

The Gata binding protein 2 (*Gata2*) belongs to an evolutionarily conserved family of C4 zinc finger transcription factors. *Gata2* was first demonstrated to be essential for hematopoiesis because *Gata2* knockout embryos die around E10 due to a failure in primitive hematopoiesis (Tsai et al. 1994). In addition to the hematopoietic lineage, a recent study systematically examined the expression of *Gata2* in ECs during embryonic development by using *Gata2*-GFP knock-in mice. GFP was strongly expressed in arterial and venous BECs (Khandekar et al. 2007). Interestingly, GFP expression was also observed in LECs budding from the veins and in postnatal lymphatic vessels (Khandekar et al. 2007). Gata2 expression was also reported in lymphatic valve cells, suggesting a possible role for the gene in lymphatic valve formation (Kazenwadel et al. 2012). Importantly, conditional inactivation of *Gata2* in the endothelial lineage led to edema and hemorrhaging and ultimately embryonic demise at around E16.5. Further analysis revealed that loss of *Gata2* caused lymph sac hypoplasia and suggested defective blood–lymphatic separation (Lim et al. 2012). More evidence for *Gata2* function in lymphatic vascular development comes from human patients in whom mutations in *Gata2* have been characterized as the cause of primary lymphedema associated with a predisposition to acute myeloid leukemia (Emberger syndrome) (Ostergaard et al. 2011). Similarly, in some patients with myelodysplastic syndrome, acute myeloid leukemia, and "MonoMAC" syndrome, mutations in *Gata2* are associated with primary lymphedema (Kazenwadel et al. 2012). Little is known about the transcriptional regulation of *Gata2*. A fragment in intron 4 of *Gata2* has been identified as an endothelium-specific enhancer of *Gata2*. Analysis of this fragment revealed that transcription factors belonging to the *Ets* family and *Scl* are activators of *Gata2* expression (Khandekar et al. 2007) (Table 2.1). In addition to the upstream regulation of *Gata2*, in vitro siRNA data suggest that *Gata2* regulates the expression of many genes required for valve formation (e.g., *Prox1*, *Foxc2*, *NFATc1*, and *Itga9*) (Kazenwadel et al. 2012). Taken together, these findings suggest that *Gata2* is another newly identified lymphatic-specific transcription factor important for early lymphatic vascular development.

Acknowledgment We want to thank Josh Stokes (St. Jude) for the generation of Fig. 2.1.

References

Aranguren, X. L., Beerens, M., Coppiello, G., Wiese, C., Vandersmissen, I., Nigro, A. L., et al. (2013). COUP-TFII orchestrates venous and lymphatic endothelial identity by homo- or heterodimerisation with PROX1. *Journal of Cell Science, 126*(Pt 5), 1164–1175.

Bos, F. L., Caunt, M., Peterson-Maduro, J., Planas-Paz, L., Kowalski, J., Karpanen, T., et al. (2011). CCBE1 is essential for mammalian lymphatic vascular development and enhances the lymphangiogenic effect of vascular endothelial growth factor-C in vivo. *Circulation Research, 109*, 486–491.

Brice, G., Mansour, S., Bell, R., Collin, J. R., Child, A. H., Brady, A. F., et al. (2002). Analysis of the phenotypic abnormalities in lymphoedema-distichiasis syndrome in 74 patients with FOXC2 mutations or linkage to 16q24. *Journal of Medical Genetics, 39*, 478–483.

Carter, T. C., & Phillips, J. S. (1954). RAGGED, a semidominant coat texture mutant in the house mouse. *Journal of Heredity, 45*, 151–154.

Casley-Smith, J. R. (1980). The fine structure and functioning of tissue channels and lymphatics. *Lymphology, 13*, 177–183.

Cermenati, S., Moleri, S., Cimbro, S., Corti, P., Del Giacco, L., Amodeo, R., et al. (2008). Sox18 and Sox7 play redundant roles in vascular development. *Blood, 111*, 2657–2666.

Chang, C. P., Neilson, J. R., Bayle, J. H., Gestwicki, J. E., Kuo, A., Stankunas, K., et al. (2004). A field of myocardial-endocardial NFAT signaling underlies heart valve morphogenesis. *Cell, 118*, 649–663.

Chen, C. Y., Bertozzi, C., Zou, Z., Yuan, L., Lee, J. S., Lu, M., et al. (2012). Blood flow reprograms lymphatic vessels to blood vessels. *Journal of Clinical Investigation, 122*, 2006–2017.

Chen, L., Mupo, A., Huynh, T., Cioffi, S., Woods, M., Jin, C., et al. (2010). Tbx1 regulates Vegfr3 and is required for lymphatic vessel development. *Journal of Cell Biology, 189*, 417–424.

Dagenais, S. L., Hartsough, R. L., Erickson, R. P., Witte, M. H., Butler, M. G., & Glover, T. W. (2004). Foxc2 is expressed in developing lymphatic vessels and other tissues associated with lymphedema-distichiasis syndrome. *Gene Expression Patterns, 4*, 611–619.

Davis, R. B., Curtis, C. D., & Griffin, C. T. (2013). BRG1 promotes COUP-TFII expression and venous specification during embryonic vascular development. *Development, 140*, 1272–1281.

de la Pompa, J. L., Timmerman, L. A., Takimoto, H., Yoshida, H., Elia, A. J., Samper, E., et al. (1998). Role of the NF-ATc transcription factor in morphogenesis of cardiac valves and septum. *Nature, 392*, 182–186.

De Val, S., & Black, B. L. (2009). Transcriptional control of endothelial cell development. *Developmental Cell, 16*, 180–195.

Deng, Y., Atri, D., Eichmann, A., & Simons, M. (2013). Endothelial ERK signaling controls lymphatic fate specification. *Journal of Clinical Investigation, 123*, 1202–1215.

Dumont, D. J., Jussila, L., Taipale, J., Lymboussaki, A., Mustonen, T., Pajusola, K., et al. (1998). Cardiovascular failure in mouse embryos deficient in VEGF receptor-3. *Science, 282*, 946–949.

Fang, J., Dagenais, S. L., Erickson, R. P., Arlt, M. F., Glynn, M. W., Gorski, J. L., et al. (2000). Mutations in FOXC2 (MFH-1), a forkhead family transcription factor, are responsible for the hereditary lymphedema-distichiasis syndrome. *American Journal of Human Genetics, 67*, 1382–1388.

Flister, M. J., Wilber, A., Hall, K. L., Iwata, C., Miyazono, K., Nisato, R. E., et al. (2009). Inflammation induces lymphangiogenesis through up-regulation of VEGFR-3 mediated by NF-kappaB and Prox1. *Blood, 115*, 418–429.

Francois, M., Caprini, A., Hosking, B., Orsenigo, F., Wilhelm, D., Browne, C., et al. (2008). Sox18 induces development of the lymphatic vasculature in mice. *Nature, 456*, 643–647.

Fritz-Six, K. L., Dunworth, W. P., Li, M., & Caron, K. M. (2008). Adrenomedullin signaling is necessary for murine lymphatic vascular development. *Journal of Clinical Investigation, 118*, 40–50.

Gridley, T. (2010). Notch signaling in the vasculature. *Developmental Biology, 92*, 277–309.

Hagerling, R., Pollmann, C., Andreas, M., Schmidt, C., Nurmi, H., Adams, R. H., et al. (2013). A novel multistep mechanism for initial lymphangiogenesis in mouse embryos based on ultramicroscopy. *EMBO Journal, 32*(5), 629–644.

Harada, K., Yamazaki, T., Iwata, C., Yoshimatsu, Y., Sase, H., Mishima, K., et al. (2009). Identification of targets of Prox1 during in vitro vascular differentiation from embryonic stem cells: Functional roles of HoxD8 in lymphangiogenesis. *Journal of Cell Science, 122*, 3923–3930.

Harvey, N. L., Srinivasan, R. S., Dillard, M. E., Johnson, N. C., Witte, M. H., Boyd, K., et al. (2005). Lymphatic vascular defects promoted by Prox1 haploinsufficiency cause adult-onset obesity. *Nature Genetics, 37*, 1072–1081.

Herpers, R., van de Kamp, E., Duckers, H. J., & Schulte-Merker, S. (2008). Redundant roles for Sox7 and Sox18 in arteriovenous specification in zebrafish. *Circulation Research, 102*, 12–15.

Hogan, B. M., Bos, F. L., Bussmann, J., Witte, M., Chi, N. C., Duckers, H. J., & Schulte-Merker, S. (2009). Ccbe1 is required for embryonic lymphangiogenesis and venous sprouting. *Nature Genetics, 41*, 396–398.

Hong, Y. K., Harvey, N., Noh, Y. H., Schacht, V., Hirakawa, S., Detmar, M., & Oliver, G. (2002). Prox1 is a master control gene in the program specifying lymphatic endothelial cell fate. *Developmental Dynamics, 225*, 351–357.

Hosking, B., Francois, M., Wilhelm, D., Orsenigo, F., Caprini, A., Svingen, T., et al. (2009). Sox7 and Sox17 are strain-specific modifiers of the lymphangiogenic defects caused by Sox18 dysfunction in mice. *Development, 136*, 2385–2391.

Ichise, T., Yoshida, N., & Ichise, H. (2010). H-, N- and Kras cooperatively regulate lymphatic vessel growth by modulating VEGFR3 expression in lymphatic endothelial cells in mice. *Development, 137*, 1003–1013.

Irrthum, A., Devriendt, K., Chitayat, D., Matthijs, G., Glade, C., Steijlen, P. M., et al. (2003). Mutations in the transcription factor gene SOX18 underlie recessive and dominant forms of hypotrichosis-lymphedema-telangiectasia. *American Journal of Human Genetics, 72*, 1470–1478.

Johnson, N. C., Dillard, M. E., Baluk, P., McDonald, D. M., Harvey, N. L., Frase, S. L., & Oliver, G. (2008). Lymphatic endothelial cell identity is reversible and its maintenance requires Prox1 activity. *Genes and Development, 22*, 3282–3291.

Kanady, J. D., Dellinger, M. T., Munger, S. J., Witte, M. H., & Simon, A. M. (2011). Connexin37 and Connexin43 deficiencies in mice disrupt lymphatic valve development and result in lymphatic disorders including lymphedema and chylothorax. *Developmental Biology, 354*, 253–266.

Kang, J., Yoo, J., Lee, S., Tang, W., Aguilar, B., Ramu, S., et al. (2010). An exquisite cross-control mechanism among endothelial cell fate regulators directs the plasticity and heterogeneity of lymphatic endothelial cells. *Blood, 116*, 140–150.

Karkkainen, M. J., Haiko, P., Sainio, K., Partanen, J., Taipale, J., Petrova, T. V., et al. (2004). Vascular endothelial growth factor C is required for sprouting of the first lymphatic vessels from embryonic veins. *Nature Immunology, 5*, 74–80.

Kazenwadel, J., Michael, M. Z., & Harvey, N. L. (2010). Prox1 expression is negatively regulated by miR-181 in endothelial cells. *Blood, 116*, 2395–2401.

Kazenwadel, J., Secker, G. A., Liu, Y. J., Rosenfeld, J. A., Wildin, R. S., Cuellar-Rodriguez, J., et al. (2012). Loss-of-function germline GATA2 mutations in patients with MDS/AML or MonoMAC syndrome and primary lymphedema reveal a key role for GATA2 in the lymphatic vasculature. *Blood, 119*, 1283–1291.

Khandekar, M., Brandt, W., Zhou, Y., Dagenais, S., Glover, T. W., Suzuki, N., et al. (2007). A Gata2 intronic enhancer confers its pan-endothelia-specific regulation. *Development, 134*, 1703–1712.

Kokubo, H., Miyagawa-Tomita, S., Nakazawa, M., Saga, Y., & Johnson, R. L. (2005). Mouse hesr1 and hesr2 genes are redundantly required to mediate Notch signaling in the developing cardiovascular system. *Developmental Biology, 278*, 301–309.

Kriederman, B. M., MyLoyde, T. L., Witte, M. H., Dagenais, S. L., Witte, C. L., Rennels, M., et al. (2003). FOXC2 haploinsufficient mice are a model for human autosomal dominant lymphedema-distichiasis syndrome. *Human Molecular Genetics, 12*, 1179–1185.

Kulkarni, R. M., Greenberg, J. M., & Akeson, A. L. (2009). NFATc1 regulates lymphatic endothelial development. *Mechanisms of Development, 126*, 350–365.

Lawson, N. D., Scheer, N., Pham, V. N., Kim, C. H., Chitnis, A. B., Campos-Ortega, J. A., et al. (2001). Notch signaling is required for arterial–venous differentiation during embryonic vascular development. *Development, 128*, 3675–3683.

Lee, S., Kang, J., Yoo, J., Ganesan, S. K., Cook, S. C., Aguilar, B., et al. (2009). Prox1 physically and functionally interacts with COUP-TFII to specify lymphatic endothelial cell fate. *Blood, 113*, 1856–1859.

Lim, K. C., Hosoya, T., Brandt, W., Ku, C. J., Hosoya-Ohmura, S., Camper, S. A., et al. (2012). Conditional Gata2 inactivation results in HSC loss and lymphatic mispatterning. *Journal of Clinical Investigation, 122*, 3705–3717.

Lin, F. J., Chen, X., Qin, J., Hong, Y. K., Tsai, M. J., & Tsai, S. Y. (2010). Direct transcriptional regulation of neuropilin-2 by COUP-TFII modulates multiple steps in murine lymphatic vessel development. *Journal of Clinical Investigation, 120*, 1694–1707.

Mansir, T., Lacombe, D., Lamireau, T., Taine, L., Chateil, J. F., Le Bail, B., et al. (1999). Abdominal lymphatic dysplasia and 22q11 microdeletion. *Genetic Counseling, 10*, 67–70.

Mishima, K., Watabe, T., Saito, A., Yoshimatsu, Y., Imaizumi, N., Masui, S., et al. (2007). Prox1 induces lymphatic endothelial differentiation via integrin alpha9 and other signaling cascades. *Molecular Biology of the Cell, 18*, 1421–1429.

Murtomaki, A., Uh, M. K., Choi, Y. K., Kitajewski, C., Borisenko, V., Kitajewski, J., & Shawber, C. J. (2013). Notch1 functions as a negative regulator of lymphatic endothelial cell differentiation in the venous endothelium. *Development, 140*(11), 2365–2376.

Niessen, K., Zhang, G., Ridgway, J. B., Chen, H., Kolumam, G., Siebel, C. W., & Yan, M. (2011). The Notch1-Dll4 signaling pathway regulates mouse postnatal lymphatic development. *Blood, 118*, 1989–1997.

Norrmen, C., Ivanov, K. I., Cheng, J., Zangger, N., Delorenzi, M., Jaquet, M., et al. (2009). FOXC2 controls formation and maturation of lymphatic collecting vessels through cooperation with NFATc1. *Journal of Cell Biology, 185*, 439–457.

Oliver, G. (2004). Lymphatic vasculature development. *Nature Reviews Immunology, 4*, 35–45.

Oliver, G., Sosa-Pineda, B., Geisendorf, S., Spana, E. P., Doe, C. Q., & Gruss, P. (1993). Prox 1, a prospero-related homeobox gene expressed during mouse development. *Mechanisms of Development, 44*, 3–16.

Ostergaard, P., Simpson, M. A., Connell, F. C., Steward, C. G., Brice, G., Woollard, W. J., et al. (2011). Mutations in GATA2 cause primary lymphedema associated with a predisposition to acute myeloid leukemia (Emberger syndrome). *Nature Genetics, 43*, 929–931.

Pan, M. R., Chang, T. M., Chang, H. C., Su, J. L., Wang, H. W., & Hung, W. C. (2009). Sumoylation of Prox1 controls its ability to induce VEGFR3 expression and lymphatic phenotypes in endothelial cells. *Journal of Cell Science, 122*, 3358–3364.

Pedrioli, D. M., Karpanen, T., Dabouras, V., Jurisic, G., van de Hoek, G., Shin, J. W., et al. (2010). miR-31 functions as a negative regulator of lymphatic vascular lineage-specific differentiation in vitro and vascular development in vivo. *Molecular and Cellular Biology, 30*, 3620–3634.

Pendeville, H., Winandy, M., Manfroid, I., Nivelles, O., Motte, P., Pasque, V., et al. (2008). Zebrafish Sox7 and Sox18 function together to control arterial-venous identity. *Developmental Biology, 317*, 405–416.

Pennisi, D., Bowles, J., Nagy, A., Muscat, G., & Koopman, P. (2000a). Mice null for sox18 are viable and display a mild coat defect. *Molecular and Cellular Biology, 20*, 9331–9336.

Pennisi, D., Gardner, J., Chambers, D., Hosking, B., Peters, J., Muscat, G., et al. (2000b). Mutations in Sox18 underlie cardiovascular and hair follicle defects in ragged mice. *Nature Genetics, 24*, 434–437.

Pereira, F. A., Qiu, Y., Tsai, M. J., & Tsai, S. Y. (1995). Chicken ovalbumin upstream promoter transcription factor (COUP-TF): expression during mouse embryogenesis. *Journal of Steroid Biochemistry and Molecular Biology, 53*, 503–508.

Petrova, T. V., Karpanen, T., Norrmen, C., Mellor, R., Tamakoshi, T., Finegold, D., et al. (2004). Defective valves and abnormal mural cell recruitment underlie lymphatic vascular failure in lymphedema distichiasis. *Nature Medicine, 10*, 974–981.

Petrova, T. V., Makinen, T., Makela, T. P., Saarela, J., Virtanen, I., Ferrell, R. E., et al. (2002). Lymphatic endothelial reprogramming of vascular endothelial cells by the Prox-1 homeobox transcription factor. *EMBO Journal, 21*, 4593–4599.

Ranger, A. M., Grusby, M. J., Hodge, M. R., Gravallese, E. M., de la Brousse, F. C., Hoey, T., et al. (1998). The transcription factor NF-ATc is essential for cardiac valve formation. *Nature, 392*, 186–190.

Roca, C., & Adams, R. H. (2007). Regulation of vascular morphogenesis by Notch signaling. *Genes and Development, 21*, 2511–2524.

Sabin, F. (1902). On the origin of the lymphatics system from the veins and the development of the lymph hearts and the thoracic duct in the pig. *The American Journal of Anatomy, 1*, 367–389.

Sabine, A., Agalarov, Y., Maby-El Hajjami, H., Jaquet, M., Hagerling, R., Pollmann, C., et al. (2012). Mechanotransduction, PROX1, and FOXC2 cooperate to control connexin37 and calcineurin during lymphatic-valve formation. *Developmental Cell, 22*, 430–445.

Sacchi, G., Weber, E., Agliano, M., Raffaelli, N., & Comparini, L. (1997). The structure of superficial lymphatics in the human thigh: Precollectors. *Anatomical Record, 247*, 53–62.

Shan, S. F., Wang, L. F., Zhai, J. W., Qin, Y., Ouyang, H. F., Kong, Y. Y., et al. (2008). Modulation of transcriptional corepressor activity of prospero-related homeobox protein (Prox1) by SUMO modification. *FEBS Letters, 582*, 3723–3728.

Shin, J. W., Min, M., Larrieu-Lahargue, F., Canron, X., Kunstfeld, R., Nguyen, L., et al. (2006). Prox1 promotes lineage-specific expression of fibroblast growth factor (FGF) receptor-3 in lymphatic endothelium: A role for FGF signaling in lymphangiogenesis. *Molecular Biology of the Cell, 17*, 576–584.

Slee, J. (1957a). The morphology and development of 'ragged'— A mutant affecting the skin and hair of the house mouse I. Adult morphology. *Journal of Genetics, 55*, 100–121.

Slee, J. (1957b). The morphology and development of ragged—A mutant affecting the skin and hair of the house mouse II. Genetics, Embryology and Gross Juvenile Morphology. *Journal of Genetics, 55*, 570–584.

Srinivasan, R. S., Dillard, M. E., Lagutin, O. V., Lin, F. J., Tsai, S., Tsai, M. J., et al. (2007). Lineage tracing demonstrates the venous origin of the mammalian lymphatic vasculature. *Genes and Development, 21*, 2422–2432.

Srinivasan, R. S., Geng, X., Yang, Y., Wang, Y., Mukatira, S., Studer, M., et al. (2010). The nuclear hormone receptor Coup-TFII is required for the initiation and early maintenance of Prox1 expression in lymphatic endothelial cells. *Genes and Development, 24*, 696–707.

Srinivasan, R. S., & Oliver, G. (2011). Prox1 dosage controls the number of lymphatic endothelial cell progenitors and the formation of the lymphovenous valves. *Genes and Development, 25*, 2187–2197.

Tsai, F. Y., Keller, G., Kuo, F. C., Weiss, M., Chen, J., Rosenblatt, M., et al. (1994). An early haematopoietic defect in mice lacking the transcription factor GATA-2. *Nature, 371*, 221–226.

van der Putte, S. C. (1975). The early development of the lymphatic system in mouse embryos. *Acta Morphologica Neerlando-Scandinavica, 13*, 245–286.

Wigle, J. T., Harvey, N., Detmar, M., Lagutina, I., Grosveld, G., Gunn, M. D., et al. (2002). An essential role for Prox1 in the induction of the lymphatic endothelial cell phenotype. *EMBO Journal, 21*, 1505–1513.

Wigle, J. T., & Oliver, G. (1999). Prox1 function is required for the development of the murine lymphatic system. *Cell, 98*, 769–778.

Xu, Y., Yuan, L., Mak, J., Pardanaud, L., Caunt, M., Kasman, I., et al. (2010). Neuropilin-2 mediates VEGF-C-induced lymphatic sprouting together with VEGFR3. *Journal of Cell Biology, 188*, 115–130.

Yagi, H., Furutani, Y., Hamada, H., Sasaki, T., Asakawa, S., Minoshima, S., et al. (2003). Role of TBX1 in human del22q11.2 syndrome. *Lancet, 362*, 1366–1373.

Yamazaki, T., Yoshimatsu, Y., Morishita, Y., Miyazono, K., & Watabe, T. (2009). COUP-TFII regulates the functions of Prox1 in lymphatic endothelial cells through direct interaction. *Genes to Cells, 14*, 425–434.

Yang, Y., Garcia-Verdugo, J. M., Soriano-Navarro, M., Srinivasan, R. S., Scallan, J. P., Singh, M. K., et al. (2012). Lymphatic endothelial progenitors bud from the cardinal vein and intersomitic vessels in mammalian embryos. *Blood, 120*(11), 2340–2348.

Yao, L. C., Baluk, P., Srinivasan, R. S., Oliver, G., & McDonald, D. M. (2012). Plasticity of button-like junctions in the endothelium of airway lymphatics in development and inflammation. *American Journal of Pathology, 180*, 2561–2575.

Yoshimatsu, Y., Yamazaki, T., Mihira, H., Itoh, T., Suehiro, J., Yuki, K., et al. (2011). Ets family members induce lymphangiogenesis through physical and functional interaction with Prox1. *Journal of Cell Science, 124*, 2753–2762.

You, L. R., Lin, F. J., Lee, C. T., Demayo, F. J., Tsai, M. J., & Tsai, S. Y. (2005). Suppression of Notch signalling by the COUP-TFII transcription factor regulates vein identity. *Nature, 435*, 98–104.

Zheng, W., Tammela, T., Yamamoto, M., Anisimov, A., Holopainen, T., Kaijalainen, S., et al. (2011). Notch restricts lymphatic vessel sprouting induced by vascular endothelial growth factor. *Blood, 118*, 1154–1162.

Chapter 3
Mechanosensing in Developing Lymphatic Vessels

Lara Planas-Paz and Eckhard Lammert

Abstract The lymphatic vasculature is responsible for fluid homeostasis, transport of immune cells, inflammatory molecules, and dietary lipids. It is composed of a network of lymphatic capillaries that drain into collecting lymphatic vessels and ultimately bring fluid back to the blood circulation. Lymphatic endothelial cells (LECs) that line lymphatic capillaries present loose overlapping intercellular junctions and anchoring filaments that support fluid drainage. When interstitial fluid accumulates within tissues, the extracellular matrix (ECM) swells and pulls the anchoring filaments. This results in opening of the LEC junctions and permits interstitial fluid uptake. The absorbed fluid is then transported within collecting lymphatic vessels, which exhibit intraluminal valves that prevent lymph backflow and smooth muscle cells that sequentially contract to propel lymph.

Mechanotransduction involves translation of mechanical stimuli into biological responses. LECs have been shown to sense and respond to changes in ECM stiffness, fluid pressure-induced cell stretch, and fluid flow-induced shear stress. How these signals influence LEC function and lymphatic vessel growth can be investigated by using different mechanotransduction assays in vitro and to some extent in vivo.

In this chapter, we will focus on the mechanical forces that regulate lymphatic vessel expansion during embryonic development and possibly secondary lymphedema. In mouse embryos, it has been recently shown that the amount of interstitial fluid determines the extent of lymphatic vessel expansion via a mechanosensory

---

L. Planas-Paz
Institute of Metabolic Physiology, Heinrich-Heine University, Universitätsstrasse 1, 40225 Düsseldorf, Germany

E. Lammert (✉)
Institute of Metabolic Physiology, Heinrich-Heine University, Universitätsstrasse 1, 40225 Düsseldorf, Germany

Paul-Langerhans-Group for Beta Cell Biology, German Diabetes Center (DDZ), Auf'm Hennekamp 65, 40225 Düsseldorf, Germany
e-mail: Lammert@uni-duesseldorf.de

F. Kiefer and S. Schulte-Merker (eds.), *Developmental Aspects of the Lymphatic Vascular System*, Advances in Anatomy, Embryology and Cell Biology 214,
DOI 10.1007/978-3-7091-1646-3_3,

complex formed by β1 integrin and vascular endothelial growth factor receptor-3 (VEGFR3). This model might as well apply to secondary lymphedema.

3.1 Entry of Interstitial Fluid into Lymphatic Capillaries

3.1.1 Interstitial Fluid Drainage

Lymphatic capillaries are blind-ended and are surrounded by a discontinuous basement membrane (Schulte-Merker et al. 2011). They are composed of LECs that exhibit oak leaf shape and discontinuous button-like junctions that overlap at cell edges (Baluk et al. 2007). LECs are tethered to the surrounding ECM in part via anchoring filaments composed of emilin-1 and fibrillin (Leak and Burke 1968a, b; Maby-El Hajjami 2008). The current model of fluid uptake by lymphatic capillaries involves pulling anchoring filaments as a consequence of ECM swelling. The overlapping LEC junctions are thereby opened, and fluid can be absorbed through these openings (Schulte-Merker et al. 2011). Nonetheless, reduction of anchoring filaments in gene-deficient mice only results in a mild phenotype (Danussi et al. 2008; Pereira et al. 1997), suggesting that alternative mechanisms of fluid uptake exist (Planas-Paz and Lammert 2013).

3.1.2 Interstitial Fluid Pressure

Around 20 % of human body mass consists of extracellular fluid. Most of this fluid is found between cells of a tissue, in the interstitium, and only a small part is transported inside vessels (Aukland and Reed 1993). Vertebrate organisms present blood vessels, which include capillaries and small- to large-caliber arteries and veins, and lymphatic vessels (Planas-Paz and Lammert 2013; Strilic et al. 2010). In general, arterial capillaries are permeable and present a higher hydrostatic than oncotic or colloid osmotic pressure. Since the net pressure inside arterial capillaries exceeds the net pressure of the interstitium, plasma fluid extravasates towards the interstitium (Planas-Paz and Lammert 2013; Levick and Michel 2010; Taylor 1981). In contrast, venous capillaries display a higher oncotic than hydrostatic pressure, and the net pressure inside them is lower than the net pressure in the interstitium. Altogether, these facts promote that venous capillaries absorb extravasated plasma fluid (Levick and Michel 2010; Taylor 1981). However, these two opposite fluid flows within capillary beds do not fully balance each other, and a small amount of plasma fluid remains within the interstitium. Lymphatic capillaries are therefore responsible for draining the remaining plasma fluid. This is possible, since the pressure inside lymphatic capillaries is usually lower than the interstitial fluid pressure (Planas-Paz and Lammert 2013; Schmid-Schonbein 1990; Swartz and Lund 2012). Lymphatic capillaries are estimated to drain 2–4 L of fluid in humans everyday, and lymphatic collecting vessels transport lymph back to the circulatory system, thus closing the fluid cycle in organisms (Makinen et al. 2007; Földi and Strössenreuther 2005).

3.2 Transport of Lymph Fluid

3.2.1 Lymph Hearts

Lymph hearts are present in amphibians and reptiles, but are not found in mammalian organisms (Hedrick et al. 2013; Jeltsch et al. 2003). Lymph hearts are composed of an inner layer of LECs and an external muscle layer covered by fibroelastic tissue (Ny et al. 2005). Stimulation of their muscle fibers via cholinergic synapses prompts lymph hearts to rhythmically contract (Greber and Schipp 1990; Drewes et al. 2007) and to transport lymph towards the venous circulation (Jeltsch et al. 2003). Destruction of amphibian lymph hearts results in animal death after 2–4 days, since there is no other way of returning interstitial fluid to the circulatory system (Jones et al. 1997).

3.2.2 Intrinsic Lymphatic Pump

Collecting lymphatic vessels gather and transport lymph from the lymphatic capillaries to the blood circulation (Makinen et al. 2007). They are characterized by the presence of continuous zipper-like junctions and continuous basement membrane, and are surrounded by a smooth muscle cell (SMC) layer (Baluk et al. 2007; Makinen et al. 2007). These features prevent lymph from leaking out of the vessels. In addition, collecting lymphatic vessels display intraluminal valves that inhibit backward lymph flow (Petrova et al. 2004; Bazigou and Makinen 2013). The segment of lymphatic vessel between two consecutive valves is called lymphangion (Bazigou and Makinen 2013).

SMCs are innervated by adrenergic, cholinergic, and peptidergic neurons that allow SMC contraction in order to pump lymph inside collecting vessels (Ohhashi et al. 2005; von der Weid and Muthuchamy 2010; Akl et al. 2011). SMC contraction mainly depends on the following factors (von der Weid and Muthuchamy 2010; Gashev and Zawieja 2010):

1. Neural and humoral factors, including adrenergic agonists, prostanoids, bradykinin, substance P, or natriuretic factors.
2. Lymph pressure and subsequent stretching of the LECs lining lymphatic vessels. Upon increased lymph pressure and LEC stretch, SMC contraction is activated, reaching a maximum pumping activity at a pressure of ~2.2–3.7 mmHg.

3. Lymph flow induces shear stress on LECs at the inner wall of lymphatic vessels, and LECs respond by releasing nitric oxide (NO) that relaxes SMCs and thus dilates the vascular lumen.

3.2.3 Extrinsic Lymphatic Pump

Besides SMC contraction and relaxation, lymph is propelled within lymphatic vessels as a consequence of the contractility of surrounding skeletal muscle, pulsation of heart and arteries, peristaltic movements, respiratory motion, and skin compression (Gashev and Zawieja 2010). For example, lymph pressure and flow are increased postprandially due to enhanced intestinal peristalsis. Similarly, lymph pressure and flow are increased during exercise, due to enhanced skeletal muscle contractions. On the other hand, lymph flow is reduced during sleep (Olszewski et al. 1977).

3.3 Mechanotransduction

3.3.1 Definition

Cells constituting tissues are embedded in a 3D microenvironment and are attached to ECM with mechanical properties (Ingber 2006). Mechanotransduction defines the ability of cells to sense changes in ECM properties and translate them into intracellular biological responses, such as cytoskeletal rearrangements, activation of signaling cascades, and changes in gene expression (Mammoto et al. 2012). As a consequence, tissues can grow or shrink, and cells can migrate or firmly adhere, differentiate, or dedifferentiate. In addition, intracellular signaling can also be transmitted to the outside to modify the external microenvironment of the cells, for example, inducing fibrillogenesis or stiffening of ECM components. Hence mechanotransduction regulates morphogenetic events during embryonic development, arteriogenesis, bone formation and turnover, wound healing, and many other developmental processes (Hoffman et al. 2011).

3.3.2 Modes

Cells respond to various types of mechanical stimuli: ECM stiffness, pressure-induced cell stretch, flow-induced shear stress, cell compression, or osmotic stress:

1. *Stiffness* is the ability of a tissue or cell to resist deformation. It is defined by the modulus of elasticity E and is mathematically defined as force applied per unit

area divided by the resultant strain (Engler et al. 2007). In the body, both soft tissues, like brain and adipose tissue, and stiff tissues, like bone, can be found. Brain is a very compliant tissue with a modulus of elasticity of 0.1–1 kPa, while muscles and collagenous bone have stiffness of 8–17 kPa and over 34 kPa, respectively (Engler et al. 2006). The differential specification of mesenchymal stem cells (MSCs) upon changes in ECM elasticity has been very well studied (Engler et al. 2006). MSCs cultivated on soft matrices are neurogenic, on medium stiff matrices are myogenic, and on rigid matrices are osteogenic (Engler et al. 2006).

2. *Pressure* is defined as force applied to a given area. Cells exposed to pressure have been found to stretch perpendicular to the applied pressure. Pressure has been associated with lung and bone development, proliferation of gastrointestinal epithelial cells, and, in Drosophila melanogaster, midgut differentiation and mesoderm invagination (Mammoto et al. 2012).
3. *Flow-induced shear stress* is defined as force that acts tangential or nearly parallel to cells. Cardiomyocytes and endothelial and hematopoietic cells are the classical examples of cells subject to shear stress, which is induced by blood flow. Flow has been shown to determine development of heart, venous, and lymphatic valves, as well as activation of hematopoietic stem cells in the aorta-gonad-mesonephron (AGM) region and blood vessel morphogenesis (Jones et al. 1997).
4. *Compression* is defined as force that reduces cell length in the direction of the applied force and has been involved in wing development in Drosophila (Chen et al. 1997).
5. *Osmosis* determines cell volume. Cells can pump ions across the plasma membrane via ion channels, and water is osmotically required to cross the membrane in the same direction. Therefore, hypoosmotic and hyperosmotic solutions induce increase or reduction in cell volume, respectively. Osmotic stress has been shown to alter gene expression, actin organization, and calcium signaling (Finan and Guilak 2010).

3.3.3 Assays

The effect of mechanical stimuli on cells can be studied by using various well-established assays:

1. Stiffness assays can be used in order to investigate the effect of substrate compliance or resistance on cells. Such an assay requires the preparation of polyacrylamide hydrogels of certain stiffness on glass coverslips. Before cells are plated on the hydrogels, an ECM component (e.g., fibronectin, laminin, collagen) is covalently coated (Tse and Engler 2010).
2. The effect of cell stretching is typically investigated by growing cells on flexible/deformable membranes, most commonly made of silicon rubber. Cells can either be elongated in a static manner or be periodically elongated over time, thus

leading to static or cyclic stretching, respectively. Stretched cells tend to align perpendicular to the direction of stretch (Richardson et al. 2011; Zhao et al. 2011).

3. Assays have been developed to mimic the effect of fluid flow-mediated shear stress on cells. Laminar flow mimics fluid movement with no mixing of fluid streams and is usually applied at 10–20 dyn/cm^2 on ECs and 4 dyn/cm^2 on LECs (Batra et al. 2012; Orr et al. 2008; Sabine et al. 2012). Oscillatory flow mimics sites of disturbed flow, like vessel bifurcations, branching points, or sites of vessel curvature. Oscillatory flow is usually generated using an infusion-withdrawal pump combined with a peristaltic pump to obtain forward flow at varying frequency (Orr et al. 2008; Sabine et al. 2012; Arnsdorf et al. 2009; Hahn et al. 2009). For comparison, blood flow and lymph flow measurements of rat mesenteric microvasculature or prenodal lymphatic vessels are shown in Table 3.1 (Dixon et al. 2006; Pries et al. 1995; Pyke and Tschakovsky 2005; le Noble et al. 2005).
4. Optical tweezers have been recently developed to manipulate proteins or DNA by exerting extremely small forces with a highly focused laser beam. The molecules are grabbed from both sides by two beads held in place by an optical trap or by suction on a micropipette or placed on a glass surface. This allows, for example, measurement of forces generated when enzymes undergo conformational changes or the analysis of molecular motor movement (e.g., kinesin movement along microtubules) (Moffitt et al. 2008; Hormeno and Arias-Gonzalez 2006).
5. Magnetic beads are coated with specific antibodies or ligands that bind a certain membrane protein and are subject to oscillating magnetic fields. The displacement of beads can be measured and reflects the stiffness of the bead-cytoskeletal linkage (Leckband et al. 2011).

3.3.4 Molecules

Transmembrane proteins like integrins, cadherins, and G protein-coupled receptors (GPCRs) have been defined as mechanosensors, since they can undergo a force-induced conformational switch (Leckband et al. 2011; Ando and Yamamoto 2009; Schwartz and DeSimone 2008; Friedland et al. 2009). Mechanosensitive channels can also be opened or closed in response to osmotic stress, stretch, or pressure. In eukaryotic cells, the following mechanosensitive channels exist: (1) excitatory cationic channels, subdivided into the Piezo family, transient receptor potential (TRP) family, and degenerin (DEG) family, and (2) inhibitory K^+-selective 2P channels (Sukharev and Sachs 2012). They mediate flux of specific ions across the plasma membrane, thus modifying the electrical potential of cells (Sukharev and Sachs 2012). The primary cilium contains mechanosensitive ion channels that mediate mechanotransduction in response to fluid flow, pressure, touch, and vibration (Hoey et al. 2012; Berbari et al. 2009).

Table 3.1 Flow and shear stress in blood and lymphatic vessels

	Blood microvasculature		
	Arterioles/arteries	Venules/veins	Prenodal lymphatic vessels
Fluid flow (mm/s)	2–8	1–2	0.9
Shear stress (dyn/cm^2)	100	5	0.4–0.6

Gap junctions are composed of six monomers of connexins and form intercellular channels that can be opened and closed by rotation of the connexin monomers upon mechanical force (Jiang et al. 2007; Salameh and Dhein 2013). Bone cells have been shown to upregulate the expression of connexin 43 (Cx43) at regions of high mechanical load, thus regulating bone cell differentiation (Jiang et al. 2007). In addition, Cx43 expression is also regulated by mechanical stimuli in endothelial cells and cardiomyocytes (Sabine et al. 2012; Salameh and Dhein 2013).

Yes-associated protein (YAP) and transcriptional co-activator with PDZ-binding motif (TAZ) were shown to translocate from cytoplasm to cell nuclei with increased ECM rigidity and to be required for mesenchymal stem cell differentiation into osteoblasts or adipocytes upon changes in ECM stiffness in vitro (Dupont et al. 2011).

Caveolae are plasma membrane invaginations particularly abundant in muscle cells, adipocytes, and ECs (Sinha et al. 2011). Since mechanical stress (e.g., osmotic swelling or unidirectional cell stretching) results in caveolae disassembly and flattening, caveolae have been proposed to rapidly buffer increases in membrane tension (Sinha et al. 2011). In addition, caveolin-1 −/− ECs display impaired responses to different types of mechanical stress (Sinha et al. 2011; Yang et al. 2011), thus supporting a role of caveolae in mechanotransduction.

3.4 Vascular Mechanotransduction

Three cell types are often found on vascular walls of large vessels, namely, vascular ECs lining the tunica intima, vascular SMCs in the tunica media, and fibroblasts in the tunica adventitia (Anwar et al. 2012). They respond to blood pressure and blood flow, which regulate morphogenesis and physiology of blood vessels (le Noble et al. 2005).

3.4.1 Blood Pressure/Stretch

Vessel walls are subject to cyclical and circumferential stretch exerted by intraluminal blood pressure (Anwar et al. 2012). Mechanical stretch in cultured SMCs has been shown to induce activation of various genes affecting vascular cell proliferation, i.e., VEGF-A, transforming growth factor-β1 (TGF-β1), hypoxia-inducible factor-1α (HIF-1α), and proliferating cell nuclear antigen (PCNA). This finding suggests that the observed hypertrophy of the tunica media in hypertensive humans (Ohya et al. 1997) and rats (Morimatsu et al. 2012; Gelosa et al. 2011) might be a consequence of increased mechanical stretch on vessel walls. Moreover, mechanical stretch also promotes the secretion of matrix metalloproteinases (MMPs) by ECs, SMCs, and fibroblasts that degrade ECM components like elastin and collagen (Intengan and Schiffrin 2001).

Under physiological conditions, human aortic arches can undergo up to 10 % expansion in diameter due to the effect of an increased blood pressure on their elastic walls (Isnard et al. 1989; O'Rourke 1995). On the contrary, the dilation of peripheral arteries (femoral, brachial, or radial arteries) is at most 5 % (Benetos et al. 1993), which might be due to a lower degree of elasticity. In hypertensive subjects, increased blood pressure can ultimately lower aortic elasticity, thus enhancing aortic stiffness and promoting arteriosclerosis (O'Rourke 1995).

3.4.2 *Blood Flow/Shear Stress*

ECs lining blood vessels sense shear stress induced by intraluminal blood flow (Hahn and Schwartz 2009). A mechanosensory complex composed of platelet and endothelial cell adhesion molecule-1 (PECAM-1), vascular endothelial cadherin (VE-cadherin), and VEGFR2 has been suggested to participate in shear stress-induced signaling in ECs (Tzima et al. 2005). The proposed model includes PECAM-1 activation after brief shear stress stimulation and subsequent activation of Src family kinases (SFKs). VE-cadherin might function as an adaptor that binds VEGFR2 through β-catenin. Tyrosine phosphorylation of VEGFR2 in response to flow requires both PECAM-1- and VE-cadherin-mediated signaling and is most likely Src-dependent. All three transmembrane proteins are required for activating αvβ3 integrins via phosphatidylinositol 3-kinase (PI3K), first at cell-cell contacts and later on the basal cell surface (Tzima et al. 2005).

Conformational activation of αvβ3 and also α5β1 integrins on ECs is followed by new binding of integrins to ECM (Tzima et al. 2001; Jalali et al. 2001). αvβ3 integrins mediate transient inactivation of small Rho GTPases that promote cyto-skeletal rearrangements and cell alignment in the direction of flow (Tzima et al. 2001). Furthermore, flow- and integrin-induced Rac1 and NF-κB activation stimulates expression of intercellular cell adhesion molecule-1 (ICAM-1), which is involved in recruiting leukocytes during atherosclerotic plaque formation (Tzima et al. 2002; Reyes-Reyes et al. 2001).

In straight vessel segments, ECs are subject to laminar flow; at branching points, vessel bifurcations, or curvatures, blood flow is rather disturbed than uniform (Hahn and Schwartz 2009; Chiu and Chien 2011). Shear stress induced by laminar flow is atheroprotective, since cells can adapt to it and downregulate its induced intracel-lular signaling. On the other hand, shear stress as a consequence of disturbed flow is atherogenic, since cells cannot adapt to constantly changing flow patterns and do not downregulate its induced signaling (Chiu and Chien 2011; Warboys et al. 2011; Weber and Noels 2011).

The requirement of PECAM-1 for mechanotransduction in response to flow in vivo is controversial, since PECAM-1 knockout mice bred in C57Bl/6 strain exhibit no obvious vascular defects (Duncan et al. 1999; Schenkel et al. 2004). However, reduced NF-κB activation and ICAM-1 expression have been reported at aortic branch points, where flow is markedly disturbed (Tzima et al. 2005). Moreover, pecam-1- and apolipoprotein E (ApoE)-double-knockout mice exhibit a reduced size of atherosclerotic lesions within the aortic arch (Harry et al. 2008), suggesting a role of PECAM-1 in translating disturbed flow-mediated signaling into pathological outcome.

3.5 Mechanotransduction in Lymphatic Development

3.5.1 Descriptive Observations

Lymphatic development starts at around E9.0–E9.5 with the differentiation of LECs from ECs of cardinal veins (Wigle and Oliver 1999; Francois et al. 2008). LECs migrate towards the dorsolateral mesenchyme and assemble the first lymphatic vessels, named jugular lymph sacs or primordial thoracic ducts (Hägerling et al. 2013). Numerous signaling pathways are involved in the differentiation, survival, and growth of LECs (Schulte-Merker et al. 2011). In addition, mechanotransduction complexes have recently been described in LECs (Planas-Paz et al. 2012; Sabine et al. 2012) (Fig. 3.1).

Interstitial fluid pressure of embryos at different developmental stages was measured at their jugular regions and observed to peak at around E12.0 (Planas-Paz et al. 2012). Furthermore, a correlation was found between the peak in interstitial fluid pressure and maximum length of LECs, tyrosine phosphorylation of VEGFR3, and LEC proliferation (Planas-Paz et al. 2012), suggesting that interstitial fluid pressure controls key parameters of lymphatic vessel expansion. When the amount of interstitial fluid was experimentally increased or decreased via addition or removal of fluid, growth of the lymphatic vasculature was enhanced or inhibited, respectively (Planas-Paz et al. 2012).

3.5.2 Role of Integrins

β1 integrins were shown to become activated upon increased interstitial fluid volume, as well as to mediate VEGFR3 tyrosine phosphorylation and LEC proliferation in response to increased interstitial fluid volume (Planas-Paz et al. 2012).

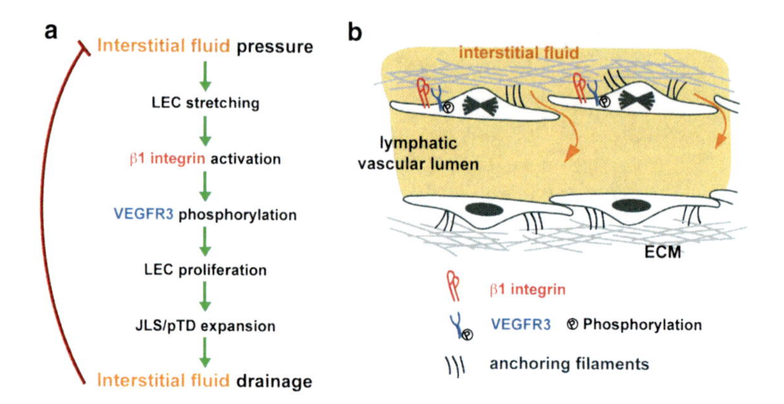

Fig. 3.1 Mechanoinduction in embryonic lymphatic vessel growth. (**a**) Model to explain the expansion of lymphatic vessels, as shown for the jugular lymph sacs (JLS), also called primordial thoracic ducts (pTD), in early mouse embryos. (**b**) Schematic diagram of fluid drainage by lymphatic vessels. Following plasma fluid accumulation in the mesenchyme, the interstitial fluid pressure increases and causes stretching of LECs attached to the swollen ECM. β1 integrin is subsequently activated in the stretched LECs and leads, possibly via activation of SFK, to VEGFR3 tyrosine phosphorylation and signaling. Therefore, LECs start to proliferate, and lymphatic vessels enlarge and more efficiently drain the accumulated interstitial fluid (Planas-Paz and Lammert 2013)

Genetic deletion of β1 integrin in ECs of mouse embryos resulted in reduced VEGFR3 tyrosine phosphorylation and reduced LEC proliferation and total LEC numbers and was accompanied by general edema. Furthermore, LECs of *β1 integrin*-deficient mice showed a blunted response to increases in interstitial fluid volume (i.e., LECs increased neither their VEGFR3 tyrosine phosphorylation nor their proliferation) (Planas-Paz et al. 2012). Therefore, β1 integrin signaling is required for translating an increased interstitial fluid volume into lymphatic vessel expansion.

To further corroborate the requirement of β1 integrins in LEC mechanotransduction, unidirectional stretching of cultured LECs was used to mimic the increased interstitial fluid pressure. β1 integrin silencing by two different siRNAs abolished the increase in VEGFR3 tyrosine phosphorylation and LEC proliferation upon LEC stretching, thus validating β1 integrins as mechanotransducers in LECs (Planas-Paz et al. 2012).

3.5.3 Role of VEGFR3

Not only β1 integrin but also VEGFR3 signaling was proven necessary in LEC mechanotransduction. Blocking VEGFR3 signaling by injecting VEGFR3-Fc fusion proteins in mouse embryos reduced both VEGFR3 tyrosine phosphorylation and LEC proliferation in response to fluid accumulation (Planas-Paz et al. 2012).

The combination of VEGF-C and mechanical stretching on LECs was shown to induce a synergistic increase in VEGFR3 tyrosine phosphorylation and LEC proliferation. In vitro VEGFR3 silencing with specific siRNAs reduced LEC proliferation upon mechanical stretching (Planas-Paz et al. 2012).

3.5.4 *Possible Role of SFKs*

β1 integrin binding to ECM can induce activation of SFK (Klinghoffer et al. 1999). In addition, SFKs were shown to phosphorylate VEGFR3 at tyrosine residues (Galvagni et al. 2010) that are different from the residues phosphorylated by VEGF-C binding to the extracellular domain of VEGFR3 (Dixelius et al. 2003). Pharmacological inhibition of SFKs abolishes ECM-induced VEGFR3 phosphorylation (Galvagni et al. 2010). Thus, SFKs are suitable candidates for mediating β1 integrin-induced VEGFR3 phosphorylation upon fluid accumulation or LEC stretch.

3.6 Mechanotransduction in Secondary Lymph Edema Formation

3.6.1 *Secondary Lymph Edema*

Lymphatic vessel dysfunction is associated with decreased fluid drainage and transport and can lead to plasma fluid accumulation in interstitial spaces and ultimately to permanent tissue swelling or lymphedema (Warren et al. 2007).

Lymphedema is classified as primary or secondary depending on the underlying mechanism. Primary lymphedema is a rare genetic disease, caused by mutations in critical lymphangiogenic genes (Warren et al. 2007). Secondary lymphedema includes 99 % of all lymphedema cases and is caused by (Warren et al. 2007; Radhakrishnan and Rockson 2008):

1. Surgical removal of lymph nodes that have been infiltrated by tumor cells
2. Radiotherapy-associated damage of lymphatic vessels
3. Filarial infection

The leading cause of secondary lymphedema in industrialized countries is breast cancer surgery, since 30 % of patients develop lymphedema of an upper limb following an axillary lymph node dissection (Warren et al. 2007). Moreover, a hyperplastic lymphatic vasculature and abnormal lymphatic valves are often associated (Mellor et al. 2000; Loprinzi et al. 1996).

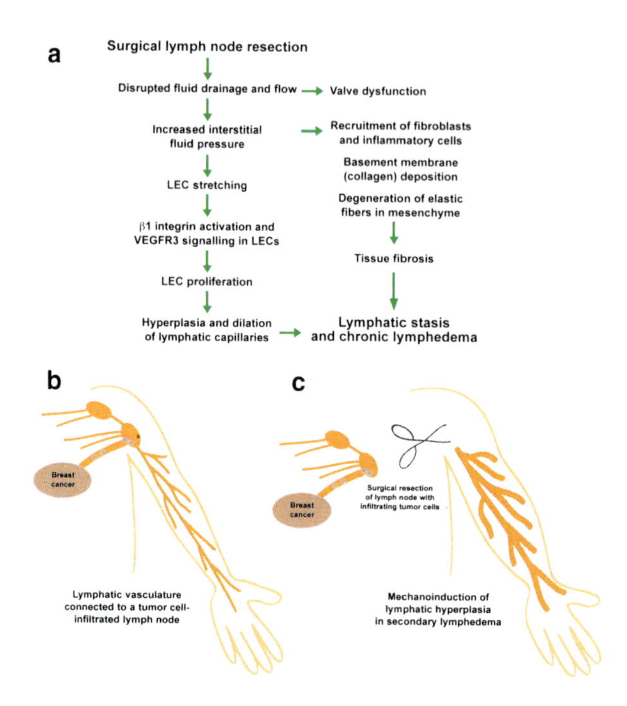

Fig. 3.2 Mechanotransduction in secondary lymphedema formation—a hypothetical model. (**a**) Hypothetical stepwise model of the possible contribution of mechanotransduction to lymphatic hyperplasia in secondary lymphedema. (**b**) Schematic diagram of a physiological lymphatic vasculature that drains interstitial fluid from an upper limb into a tumor cell-infiltrated lymph node of a breast cancer patient. (**c**) After surgical lymph node removal, lymph flow within the lymphatic vasculature is impaired, preventing the lymphatic capillaries from properly draining interstitial fluid. Consequently, the fluid pressure in the interstitium rises and triggers lymphatic vessel hyperplasia (Planas-Paz and Lammert 2013)

3.6.2 Possible Mechanisms

Surgical dissection of tumor cell-infiltrated lymph nodes interrupts lymph transport between lymphatic vessels upstream (afferent) and downstream (efferent) of the removed lymph nodes. Since lymph from the afferent side cannot be transported away, it accumulates within lymphatic vessels, which might lead to vessel dilation and reduced lymph drainage. A hypothetical consequence of the subsequent accumulation of interstitial fluid in the non-drained surrounding tissue might include an increase in interstitial fluid pressure and swelling of ECM to which LECs are attached. Following this hypothesis, in the stretched LECs, β1 integrin activation and VEGFR3 signaling could promote LEC proliferation and lymphatic vessel hyperplasia (Planas-Paz and Lammert 2013) (Fig. 3.2).

3.6.3 *Palliation of Lymph Edema*

Current palliative treatments for lymphedema are based on the use of compressive garments and lymphatic massage, which together achieve a decrease in limb volume of an affected patient by 40–60 % (Warren et al. 2007). However, these are not curative treatments, and further analyses are required to investigate whether directing lymphatic massage towards the region that requires lymphatic vessel regeneration will reduce lymphedema formation (Planas-Paz and Lammert 2013).

3.7 Conclusions/Outlook

LECs have the ability to sense changes in the amount of interstitial fluid and to thereby control lymphatic vessel expansion. β1 integrin and VEGFR3 signaling have been found to be required for mechanoinduced LEC proliferation. Additional research on how these pathways synergistically regulate lymphatic growth in disease could prove valuable for efficient treatment in the future.

References

Akl, T. J., Nagai, T., Cote, G. L., & Gashev, A. A. (2011). Mesenteric lymph flow in adult and aged rats. *American Journal of Physiology. Heart and Circulatory Physiology, 301*(5), H1828–H1840. doi:10.1152/ajpheart.00538.2011.

Ando, J., & Yamamoto, K. (2009). Vascular mechanobiology: Endothelial cell responses to fluid shear stress. *Circulation Journal, 73*(11), 1983–1992.

Anwar, M. A., Shalhoub, J., Lim, C. S., Gohel, M. S., & Davies, A. H. (2012). The effect of pressure-induced mechanical stretch on vascular wall differential gene expression. *Journal of Vascular Research, 49*(6), 463–478. doi:10.1159/000339151.

Arnsdorf, E. J., Tummala, P., Kwon, R. Y., & Jacobs, C. R. (2009). Mechanically induced osteogenic differentiation–the role of RhoA, ROCKII and cytoskeletal dynamics. *Journal of Cell Science, 122*(Pt 4), 546–553. doi:10.1242/jcs.036293.

Aukland, K., & Reed, R. K. (1993). Interstitial-lymphatic mechanisms in the control of extracellular fluid volume. *Physiological Reviews, 73*(1), 1–78.

Baluk, P., Fuxe, J., Hashizume, H., Romano, T., Lashnits, E., Butz, S., et al. (2007). Functionally specialized junctions between endothelial cells of lymphatic vessels. *Journal of Experimental Medicine, 204*(10), 2349–2362. doi:10.1084/jem.20062596.

Batra, N., Burra, S., Siller-Jackson, A. J., Gu, S., Xia, X., Weber, G. F., et al. (2012). Mechanical stress-activated integrin alpha5beta1 induces opening of connexin 43 hemichannels. *Proceedings of the National Academy of Sciences of the United States of America, 109*(9), 3359–3364. doi:10.1073/pnas.1115967109.

Bazigou, E., & Makinen, T. (2013). Flow control in our vessels: Vascular valves make sure there is no way back. *Cellular and Molecular Life Sciences, 70*(6), 1055–1066. doi:10.1007/s00018-012-1110-6.

Benetos, A., Laurent, S., Hoeks, A. P., Boutouyrie, P. H., & Safar, M. E. (1993). Arterial alterations with aging and high blood pressure. A noninvasive study of carotid and femoral arteries. *Arteriosclerosis and Thrombosis, 13*(1), 90–97.

Berbari, N. F., O'Connor, A. K., Haycraft, C. J., & Yoder, B. K. (2009). The primary cilium as a complex signaling center. *Current Biology, 19*(13), R526–R535. doi:10.1016/j.cub.2009.05.025.

Chen, C. S., Mrksich, M., Huang, S., Whitesides, G. M., & Ingber, D. E. (1997). Geometric control of cell life and death. *Science, 276*(5317), 1425–1428.

Chiu, J. J., & Chien, S. (2011). Effects of disturbed flow on vascular endothelium: Pathophysiological basis and clinical perspectives. *Physiological Reviews, 91*(1), 327–387. doi:10.1152/physrev.00047.2009.

Danussi, C., Spessotto, P., Petrucco, A., Wassermann, B., Sabatelli, P., Montesi, M., et al. (2008). Emilin1 deficiency causes structural and functional defects of lymphatic vasculature. *Molecular and Cellular Biology, 28*(12), 4026–4039. doi:10.1128/MCB.02062-07.

Dixelius, J., Makinen, T., Wirzenius, M., Karkkainen, M. J., Wernstedt, C., Alitalo, K., & Claesson-Welsh, L. (2003). Ligand-induced vascular endothelial growth factor receptor-3 (VEGFR-3) heterodimerization with VEGFR-2 in primary lymphatic endothelial cells regulates tyrosine phosphorylation sites. *The Journal of Biological Chemistry, 278*(42), 40973–40979. doi:10.1074/jbc.M304499200.

Dixon, J. B., Greiner, S. T., Gashev, A. A., Cote, G. L., Moore, J. E., & Zawieja, D. C. (2006). Lymph flow, shear stress, and lymphocyte velocity in rat mesenteric prenodal lymphatics. *Microcirculation, 13*(7), 597–610. doi:10.1080/10739680600893909.

Drewes, R. C., Hedrick, M. S., Hillman, S. S., & Withers, P. C. (2007). Unique role of skeletal muscle contraction in vertical lymph movement in anurans. *Journal of Experimental Biology, 210*(Pt 22), 3931–3939. doi:10.1242/jeb.009548.

Duncan, G. S., Andrew, D. P., Takimoto, H., Kaufman, S. A., Yoshida, H., Spellberg, J., et al. (1999). Genetic evidence for functional redundancy of Platelet/Endothelial cell adhesion molecule-1 (PECAM-1): CD31-deficient mice reveal PECAM-1-dependent and PECAM-1-independent functions. *Journal of Immunology, 162*(5), 3022–3030.

Dupont, S., Morsut, L., Aragona, M., Enzo, E., Giulitti, S., Cordenonsi, M., et al. (2011). Role of YAP/TAZ in mechanotransduction. *Nature, 474*(7350), 179–183. doi:10.1038/nature10137.

Engler, A. J., Rehfeldt, F., Sen, S., & Discher, D. E. (2007). Microtissue elasticity: Measurements by atomic force microscopy and its influence on cell differentiation. *Methods in Cell Biology, 83*, 521–545. doi:10.1016/S0091-679X(07)83022-6.

Engler, A. J., Sen, S., Sweeney, H. L., & Discher, D. E. (2006). Matrix elasticity directs stem cell lineage specification. *Cell, 126*(4), 677–689. doi:10.1016/j.cell.2006.06.044.

Finan, J. D., & Guilak, F. (2010). The effects of osmotic stress on the structure and function of the cell nucleus. *Journal of Cellular Biochemistry, 109*(3), 460–467. doi:10.1002/jcb.22437.

Földi, M., & Strössenreuther, R. (2005). *Foundations of manual lymph drainage* (3rd ed.). New York: Elsevier.

Francois, M., Caprini, A., Hosking, B., Orsenigo, F., Wilhelm, D., Browne, C., et al. (2008). Sox18 induces development of the lymphatic vasculature in mice. *Nature, 456*(7222), 643–647. doi:10.1038/nature07391.

Friedland, J. C., Lee, M. H., & Boettiger, D. (2009). Mechanically activated integrin switch controls alpha5beta1 function. *Science, 323*(5914), 642–644. doi:10.1126/science.1168441.

Galvagni, F., Pennacchini, S., Salameh, A., Rocchigiani, M., Neri, F., Orlandini, M., et al. (2010). Endothelial cell adhesion to the extracellular matrix induces c-Src-dependent VEGFR-3 phosphorylation without the activation of the receptor intrinsic kinase activity. *Circulation Research, 106*(12), 1839–1848. doi:10.1161/CIRCRESAHA.109.206326.

Gashev, A. A., & Zawieja, D. C. (2010). Hydrodynamic regulation of lymphatic transport and the impact of aging. *Pathophysiology, 17*(4), 277–287. doi:10.1016/j.pathophys.2009.09.002.

Gelosa, P., Sevin, G., Pignieri, A., Budelli, S., Castiglioni, L., Blanc-Guillemaud, V., et al. (2011). Terutroban, a thromboxane/prostaglandin endoperoxide receptor antagonist, prevents hypertensive vascular hypertrophy and fibrosis. *American Journal of Physiology. Heart and Circulatory Physiology, 300*(3), H762–H768. doi:10.1152/ajpheart.00880.2010.

Greber, K., & Schipp, R. (1990). Early development and myogenesis of the posterior anuran lymph hearts. *Anatomy and Embryology, 181*(1), 75–82.

Hägerling, R., Pollmann, C., Andreas, M., Schmidt, C., Nurmi, H., Adams, R. H., et al. (2013). A novel multistep mechanism for initial lymphangiogenesis in mouse embryos based on ultramicroscopy. *EMBO Journal, 32*(5), 629–644. doi:10.1038/emboj.2012.340.

Hahn, C., Orr, A. W., Sanders, J. M., Jhaveri, K. A., & Schwartz, M. A. (2009). The subendothelial extracellular matrix modulates JNK activation by flow. *Circulation Research, 104*(8), 995–1003. doi:10.1161/CIRCRESAHA.108.186486.

Hahn, C., & Schwartz, M. A. (2009). Mechanotransduction in vascular physiology and atherogenesis. *Nature Reviews Molecular Cell Biology, 10*(1), 53–62. doi:10.1038/nrm2596.

Harry, B. L., Sanders, J. M., Feaver, R. E., Lansey, M., Deem, T. L., Zarbock, A., et al. (2008). Endothelial cell PECAM-1 promotes atherosclerotic lesions in areas of disturbed flow in ApoE-deficient mice. *Arteriosclerosis, Thrombosis, and Vascular Biology, 28*(11), 2003–2008. doi:10.1161/ATVBAHA.108.164707.

Hedrick, M. S., Hillman, S. S., Drewes, R. C., & Withers, P. C. (2013). Lymphatic regulation in non-mammalian vertebrates. *Journal of Applied Physiology, 115*(3), 297–308. doi:10.1152/japplphysiol.00201.2013.

Hoey, D. A., Downs, M. E., & Jacobs, C. R. (2012). The mechanics of the primary cilium: An intricate structure with complex function. *Journal of Biomechanics, 45*(1), 17–26. doi:10.1016/j.jbiomech.2011.08.008.

Hoffman, B. D., Grashoff, C., & Schwartz, M. A. (2011). Dynamic molecular processes mediate cellular mechanotransduction. *Nature, 475*(7356), 316–323. doi:10.1038/nature10316.

Hormeno, S., & Arias-Gonzalez, J. R. (2006). Exploring mechanochemical processes in the cell with optical tweezers. *Biology of the Cell, 98*(12), 679–695. doi:10.1042/BC20060036.

Ingber, D. E. (2006). Cellular mechanotransduction: Putting all the pieces together again. *FASEB Journal, 20*(7), 811–827. doi:10.1096/fj.05-5424rev.

Intengan, H. D., & Schiffrin, E. L. (2001). Vascular remodeling in hypertension: Roles of apoptosis, inflammation, and fibrosis. *Hypertension, 38*(3 Pt 2), 581–587.

Isnard, R. N., Pannier, B. M., Laurent, S., London, G. M., Diebold, B., & Safar, M. E. (1989). Pulsatile diameter and elastic modulus of the aortic arch in essential hypertension: A noninvasive study. *Journal of the American College of Cardiology, 13*(2), 399–405.

Jalali, S., del Pozo, M. A., Chen, K., Miao, H., Li, Y., Schwartz, M. A., et al. (2001). Integrin-mediated mechanotransduction requires its dynamic interaction with specific extracellular matrix (ECM) ligands. *Proceedings of the National Academy of Sciences of the United States of America, 98*(3), 1042–1046. doi:10.1073/pnas.031562998.

Jeltsch, M., Tammela, T., Alitalo, K., & Wilting, J. (2003). Genesis and pathogenesis of lymphatic vessels. *Cell and Tissue Research, 314*(1), 69–84. doi:10.1007/s00441-003-0777-2.

Jiang, J. X., Siller-Jackson, A. J., & Burra, S. (2007). Roles of gap junctions and hemichannels in bone cell functions and in signal transmission of mechanical stress. *Frontiers in Bioscience, 12*, 1450–1462.

Jones, J. M., Gamperl, A. K., Farrell, A. P., & Toews, D. P. (1997). Direct measurement of flow from the posterior lymph hearts of hydrated and dehydrated toads (Bufo marinus). *Journal of Experimental Biology, 200*(Pt 11), 1695–1702.

Klinghoffer, R. A., Sachsenmaier, C., Cooper, J. A., & Soriano, P. (1999). Src family kinases are required for integrin but not PDGFR signal transduction. *EMBO Journal, 18*(9), 2459–2471. doi:10.1093/emboj/18.9.2459.

le Noble, F., Fleury, V., Pries, A., Corvol, P., Eichmann, A., & Reneman, R. S. (2005). Control of arterial branching morphogenesis in embryogenesis: Go with the flow. *Cardiovascular Research, 65*(3), 619–628. doi:10.1016/j.cardiores.2004.09.018.

Leak, L. V., & Burke, J. F. (1968a). Ultrastructural studies on the lymphatic anchoring filaments. *Journal of Cell Biology, 36*(1), 129–149.

Leak, L. V., & Burke, J. F. (1968b). Electron microscopic study of lymphatic capillaries in the removal of connective tissue fluids and particulate substances. *Lymphology, 1*(2), 39–52.

Leckband, D. E., le Duc, Q., Wang, N., & de Rooij, J. (2011). Mechanotransduction at cadherin-mediated adhesions. *Current Opinion in Cell Biology, 23*(5), 523–530. doi:10.1016/j.ceb.2011.08.003.

Levick, J. R., & Michel, C. C. (2010). Microvascular fluid exchange and the revised Starling principle. *Cardiovascular Research, 87*(2), 198–210. doi:10.1093/cvr/cvq062.

Loprinzi, C. L., Okuno, S., Pisansky, T. M., Sterioff, S., Gaffey, T. A., & Morton, R. F. (1996). Postsurgical changes of the breast that mimic inflammatory breast carcinoma. *Mayo Clinic Proceedings, 71*(6), 552–555. doi:10.1016/S0025-6196(11)64111-6.

Maby-El Hajjami, H. P. T. (2008). Developmental and pathological lymphangiogenesis: From models to human disease. *Histochemistry and Cell Biology, 130*(6), 1063–1078. doi:10.1007/s00418-008-0525-5.

Makinen, T., Norrmen, C., & Petrova, T. V. (2007). Molecular mechanisms of lymphatic vascular development. *Cellular and Molecular Life Sciences, 64*(15), 1915–1929. doi:10.1007/s00018-007-7040-z.

Mammoto, A., Mammoto, T., & Ingber, D. E. (2012). Mechanosensitive mechanisms in transcriptional regulation. *Journal of Cell Science, 125*(Pt 13), 3061–3073. doi:10.1242/jcs.093005.

Mellor, R. H., Stanton, A. W., Azarbod, P., Sherman, M. D., Levick, J. R., & Mortimer, P. S. (2000). Enhanced cutaneous lymphatic network in the forearms of women with postmastectomy oedema. *Journal of Vascular Research, 37*(6), 501–512. doi:10.1159/000054083.

Moffitt, J. R., Chemla, Y. R., Smith, S. B., & Bustamante, C. (2008). Recent advances in optical tweezers. *Annual Review of Biochemistry, 77*, 205–228. doi:10.1146/annurev.biochem.77.043007.090225.

Morimatsu, Y., Sakashita, N., Komohara, Y., Ohnishi, K., Masuda, H., Dahan, D., et al. (2012). Development and characterization of an animal model of severe pulmonary arterial hypertension. *Journal of Vascular Research, 49*(1), 33–42. doi:10.1159/000329594.

Ny, A., Koch, M., Schneider, M., Neven, E., Tong, R. T., Maity, S., et al. (2005). A genetic Xenopus laevis tadpole model to study lymphangiogenesis. *Nature Medicine, 11*(9), 998–1004. doi:10.1038/nm1285.

O'Rourke, M. (1995). Mechanical principles in arterial disease. *Hypertension, 26*(1), 2–9.

Ohhashi, T., Mizuno, R., Ikomi, F., & Kawai, Y. (2005). Current topics of physiology and pharmacology in the lymphatic system. *Pharmacology & Therapeutics, 105*(2), 165–188. doi:10.1016/j.pharmthera.2004.10.009.

Ohya, Y., Abe, I., Fujii, K., Kobayashi, K., Onaka, U., & Fujishima, M. (1997). Intima-media thickness of the carotid artery in hypertensive subjects and hypertrophic cardiomyopathy patients. *Hypertension, 29*(1 Pt 2), 361–365.

Olszewski, W. E. A., Jaeger, P. M., Sokolowski, J., & Theodorsen, L. (1977). Flow and composition of leg lymph in normal men during venous stasis, muscular activity and local hyperthermia. *Acta Physiologica Scandinavica, 99*(2), 149–155.

Orr, A. W., Hahn, C., Blackman, B. R., & Schwartz, M. A. (2008). p21-activated kinase signaling regulates oxidant-dependent NF-kappa B activation by flow. *Circulation Research, 103*(6), 671–679. doi:10.1161/CIRCRESAHA.108.182097.

Pereira, L., Andrikopoulos, K., Tian, J., Lee, S. Y., Keene, D. R., Ono, R., et al. (1997). Targetting of the gene encoding fibrillin-1 recapitulates the vascular aspect of Marfan syndrome. *Nature Genetics, 17*(2), 218–222. doi:10.1038/ng1097-218.

Petrova, T. V., Karpanen, T., Norrmen, C., Mellor, R., Tamakoshi, T., Finegold, D., et al. (2004). Defective valves and abnormal mural cell recruitment underlie lymphatic vascular failure in lymphedema distichiasis. *Nature Medicine, 10*(9), 974–981. doi:10.1038/nm1094.

Planas-Paz, L., & Lammert, E. (2013). Mechanical forces in lymphatic vascular development and disease. *Cellular and Molecular Life Sciences*. doi:10.1007/s00018-013-1358-5.

Planas-Paz, L., Strilic, B., Goedecke, A., Breier, G., Fässler, R., & Lammert, E. (2012). Mechanoinduction of lymph vessel expansion. *EMBO Journal, 31*(4), 788–804. doi:10.1038/emboj.2011.456.

Pries, A. R., Secomb, T. W., & Gaehtgens, P. (1995). Design principles of vascular beds. *Circulation Research, 77*(5), 1017–1023.

Pyke, K. E., & Tschakovsky, M. E. (2005). The relationship between shear stress and flow-mediated dilatation: Implications for the assessment of endothelial function. *Journal of Physiology, 568*(Pt 2), 357–369. doi:10.1113/jphysiol.2005.089755.

Radhakrishnan, K., & Rockson, S. G. (2008). The clinical spectrum of lymphatic disease. *Annals of the New York Academy of Sciences, 1131*, 155–184. doi:10.1196/annals.1413.015.

Reyes-Reyes, M., Mora, N., Zentella, A., & Rosales, C. (2001). Phosphatidylinositol 3-kinase mediates integrin-dependent NF-kappaB and MAPK activation through separate signaling pathways. *Journal of Cell Science, 114*(Pt 8), 1579–1589.

Richardson, W. J., Metz, R. P., Moreno, M. R., Wilson, E., & Moore, J. E., Jr. (2011). A device to study the effects of stretch gradients on cell behavior. *Journal of Biomechanical Engineering, 133*(10), 101008. doi:10.1115/1.4005251.

Sabine, A., Agalarov, Y., Maby-El Hajjami, H., Jaquet, M., Hägerling, R., Pollmann, C., et al. (2012). Mechanotransduction, PROX1, and FOXC2 cooperate to control connexin37 and calcineurin during lymphatic-valve formation. *Developmental Cell, 22*(2), 430–445. doi:10.1016/j.devcel.2011.12.020.

Salameh, A., & Dhein, S. (2013). Effects of mechanical forces and stretch on intercellular gap junction coupling. *Biochimica et Biophysica Acta, 1828*(1), 147–156. doi:10.1016/j.bbamem.2011.12.030.

Schenkel, A. R., Chew, T. W., & Muller, W. A. (2004). Platelet endothelial cell adhesion molecule deficiency or blockade significantly reduces leukocyte emigration in a majority of mouse strains. *Journal of Immunology, 173*(10), 6403–6408.

Schmid-Schonbein, G. W. (1990). Microlymphatics and lymph flow. *Physiological Reviews, 70*(4), 987–1028.

Schulte-Merker, S., Sabine, A., & Petrova, T. V. (2011). Lymphatic vascular morphogenesis in development, physiology, and disease. *Journal of Cell Biology, 193*(4), 607–618. doi:10.1083/jcb.201012094.

Schwartz, M. A., & DeSimone, D. W. (2008). Cell adhesion receptors in mechanotransduction. *Current Opinion in Cell Biology, 20*(5), 551–556. doi:10.1016/j.ceb.2008.05.005.

Sinha, B., Koster, D., Ruez, R., Gonnord, P., Bastiani, M., Abankwa, D., et al. (2011). Cells respond to mechanical stress by rapid disassembly of caveolae. *Cell, 144*(3), 402–413. doi:10.1016/j.cell.2010.12.031.

Strilic, B., Kucera, T., & Lammert, E. (2010). Formation of cardiovascular tubes in invertebrates and vertebrates. *Cellular and Molecular Life Sciences, 67*(19), 3209–3218. doi:10.1007/s00018-010-0400-0.

Sukharev, S., & Sachs, F. (2012). Molecular force transduction by ion channels: Diversity and unifying principles. *Journal of Cell Science, 125*(Pt 13), 3075–3083. doi:10.1242/jcs.092353.

Swartz, M. A., & Lund, A. W. (2012). Lymphatic and interstitial flow in the tumour microenvironment: Linking mechanobiology with immunity. *Nature Reviews Cancer, 12*(3), 210–219. doi:10.1038/nrc3186.

Taylor, A. E. (1981). Capillary fluid filtration. Starling forces and lymph flow. *Circulation Research, 49*(3), 557–575.

Tse, J. R., & Engler, A. J. (2010). Preparation of hydrogel substrates with tunable mechanical properties. *Current Protocols in Cell Biology*, Chapter 10:Unit 10.16. doi:10.1002/0471143030.cb1016s47.

Tzima, E., Del Pozo, M. A., Kiosses, W. B., Mohamed, S. A., Li, S., Chien, S., & Schwartz, M. A. (2002). Activation of Rac1 by shear stress in endothelial cells mediates both cytoskeletal reorganization and effects on gene expression. *EMBO Journal, 21*(24), 6791–6800.

Tzima, E., del Pozo, M. A., Shattil, S. J., Chien, S., & Schwartz, M. A. (2001). Activation of integrins in endothelial cells by fluid shear stress mediates Rho-dependent cytoskeletal alignment. *EMBO Journal, 20*(17), 4639–4647. doi:10.1093/emboj/20.17.4639.

Tzima, E., Irani-Tehrani, M., Kiosses, W. B., Dejana, E., Schultz, D. A., Engelhardt, B., et al. (2005). A mechanosensory complex that mediates the endothelial cell response to fluid shear stress. *Nature, 437*(7057), 426–431. doi:10.1038/nature03952.

von der Weid, P. Y., & Muthuchamy, M. (2010). Regulatory mechanisms in lymphatic vessel contraction under normal and inflammatory conditions. *Pathophysiology, 17*(4), 263–276. doi:10.1016/j.pathophys.2009.10.005.

Warboys, C. M., Amini, N., de Luca, A., & Evans, P. C. (2011). The role of blood flow in determining the sites of atherosclerotic plaques. *F1000 Medicine Reports, 3*, 5. doi:10.3410/M3-5.

Warren, A. G., Brorson, H., Borud, L. J., & Slavin, S. A. (2007). Lymphedema: A comprehensive review. *Annals of Plastic Surgery, 59*(4), 464–472. doi:10.1097/01.sap.0000257149.42922.7e.

Weber, C., & Noels, H. (2011). Atherosclerosis: Current pathogenesis and therapeutic options. *Nature Medicine, 17*(11), 1410–1422. doi:10.1038/nm.2538.

Wigle, J. T., & Oliver, G. (1999). Prox1 function is required for the development of the murine lymphatic system. *Cell, 98*(6), 769–778.

Yang, B., Radel, C., Hughes, D., Kelemen, S., & Rizzo, V. (2011). p190 RhoGTPase-activating protein links the beta1 integrin/caveolin-1 mechanosignaling complex to RhoA and actin remodeling. *Arteriosclerosis, Thrombosis, and Vascular Biology, 31*(2), 376–383. doi:10.1161/ATVBAHA.110.217794.

Zhao, L., Sang, C., Yang, C., & Zhuang, F. (2011). Effects of stress fiber contractility on uniaxial stretch guiding mitosis orientation and stress fiber alignment. *Journal of Biomechanics, 44*(13), 2388–2394. doi:10.1016/j.jbiomech.2011.06.033.

Chapter 4
Plasticity of Airway Lymphatics in Development and Disease

Li-Chin Yao and Donald M. McDonald

Abstract The dynamic nature of lymphatic vessels is reflected by structural and functional modifications that coincide with changes in their environment. Lymphatics in the respiratory tract undergo rapid changes around birth, during adaptation to air breathing, when lymphatic endothelial cells develop button-like intercellular junctions specialized for efficient fluid uptake and transport. In inflammatory conditions, lymphatic vessels proliferate and undergo remodeling to accommodate greater plasma leakage and immune cell trafficking. However, the newly formed lymphatics are abnormal, and resolution of inflammation is not accompanied by complete reversal of the lymphatic vessel changes back to the baseline. As the understanding of lymphatic plasticity advances, approaches for eliminating the abnormal vessels and improving the functionality of those that remain move closer to reality. This chapter provides an overview of what is known about lymphatic vessel growth, remodeling, and other forms of plasticity that occur during development or inflammation, with an emphasis on the respiratory tract. Also addressed is the limited reversibility of changes in lymphatics during the resolution of inflammation.

4.1 Introduction

Plasma leakage, edema, and remodeling of the airway wall are hallmarks of inflammatory airway diseases (Dunnill 1960; Ebina 2008; Wilson and Hii 2006). Lymphangiogenesis and lymphatic remodeling are among the features of sustained respiratory inflammation (El-Chemaly et al. 2008). Lymphatics proliferate in pneumonia (Mandal et al. 2008; Parra et al. 2012), regress in asthma (Ebina et al. 2010), and undergo remodeling and growth in idiopathic pulmonary fibrosis (Yamashita

L.-C. Yao • D.M. McDonald (✉)
Department of Anatomy, Comprehensive Cancer Center, Cardiovascular Research Institute, University of California-San Francisco, San Francisco, CA 94143, USA
e-mail: donald.mcdonald@ucsf.edu

F. Kiefer and S. Schulte-Merker (eds.), *Developmental Aspects of the Lymphatic Vascular System*, Advances in Anatomy, Embryology and Cell Biology 214,
DOI 10.1007/978-3-7091-1646-3_4, © Springer-Verlag Wien 2014

et al. 2009; El-Chemaly et al. 2009). Understanding the contribution of lymphatic changes to disease pathophysiology and the clinical implications is still at an early stage. Elucidation of the causes, consequences, and reversibility of changes in airway lymphatics will offer new therapeutic targets and treatment strategies.

In a mouse model of sustained inflammation associated with infection by the respiratory pathogen *Mycoplasma pulmonis*, lymphatics in the airways and lung undergo rapid proliferation and remodeling (Baluk et al. 2005; McDonald 2008). Lymphatic growth and remodeling typically occur together but represent different vascular responses, and the driving factors and consequences are likely to be different. As diseases worsen, the lymphatic microvasculature undergoes progressive changes in structure and function. By comparison, lymphatic growth and remodeling during development are a natural process.

The pulmonary lymphatic network arises by sprouting lymphangiogenesis from lymphatic precursor channels and sacs that initially arise from the cardinal vein. The segmental pattern of tracheal lymphatics is established before birth, but maturation of lymphatics evidenced by specialized button-like intercellular junctions occurs postnatally. Studies of changes in lymphatic junctions around birth have provided a better understanding of lymphatic dynamics during development. Lymphatic changes found in inflammation recapitulate aspects of physiological growth and remodeling in perinatal mice (Yao et al. 2012). Understanding the plasticity of lymphatics under normal conditions and in disease, along with the underlying mechanisms, is necessary for identifying new therapeutic strategies and developing new diagnostic procedures.

Most studies of lymphangiogenesis have focused on preventing rather than reversing changes in lymphatics. In prevention studies, treatment with growth factor inhibitors starts with or before the lymphangiogenic stimulus. This approach can test the involvement of factors that promote lymphangiogenesis and elucidate the contribution of lymphatic changes to disease. In reversal studies, lymphatic changes are established before the onset of treatment. This design can be used to detect factors that are responsible for the maintenance of newly formed lymphatics, develop strategies for eliminating dysfunctional lymphatics, and elucidate the consequences of lymphatic reversal in relation to disease severity. Reversal of lymphatic changes to the baseline state could be clinically important, but the understanding of this reversibility is still at an early stage.

4.2 Structure of Lymphatic Vasculature in Normal Airways

4.2.1 Segmental Arrangement of Initial Lymphatics

The trachea is the largest airway in mice but is similar in size to small bronchioles in humans. When compared to the lung, the trachea is anatomically simpler and easier

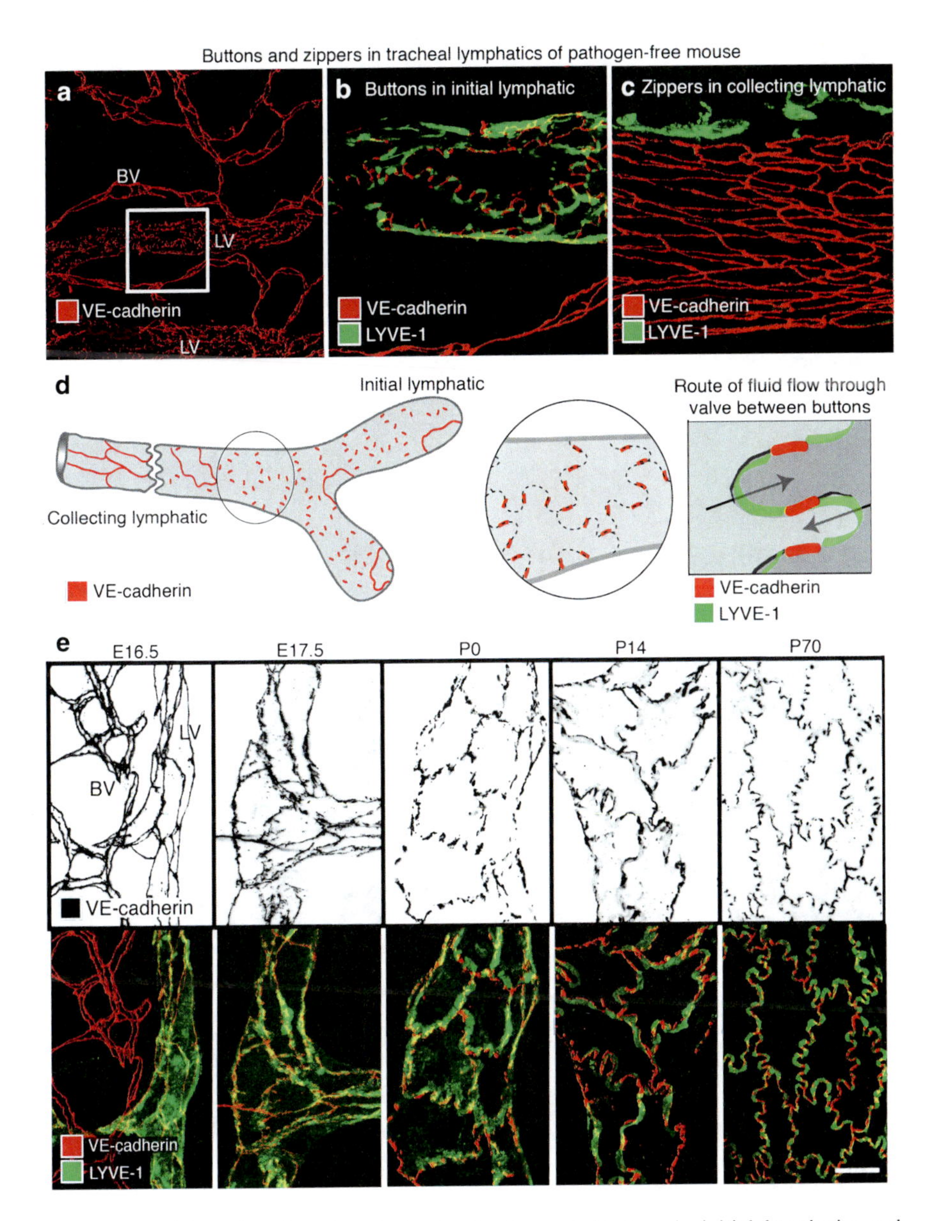

Fig. 4.1 Plasticity of tracheal lymphatics in development. Buttons in initial lymphatics and zippers in collecting lymphatics of a normal mouse trachea. (**a**) Low magnification of VE-cadherin (*red*) stained button-like junctions in initial lymphatics (LV) and zipper-like junctions in blood vessels (BV). (**b**) The *box* in (**a**) is enlarged here to show discontinuous segments of VE-cadherin immunoreactivity (*red*) at buttons and segments of LYVE-1 staining (*green*) between buttons in an initial lymphatic. (**c**) VE-cadherin immunoreactivity (*red*) at zippers in the endothelium of a collecting lymphatic with little or no LYVE-1 staining. LYVE-1-positive leukocytes (*green*) are present outside the lymphatic. (**d**) Schematic diagram showing buttons and zipper in lymphatics revealed by VE-cadherin immunoreactivity. *Middle panel* shows the oak leaf-shaped endothelial cells marked by *dashed lines*. The *right panel* shows the enlarged diagram of buttons (*red*) at the sides of cell border flaps (*green*). Fluid is believed to flow through the junction-free

to access experimentally. Moreover, the stereotypic architecture of tracheal lymphatics has proven advantageous for studying growth and remodeling of lymphatics in mouse models. The distinctive segmented arrangement of blood vessels and lymphatics is evident in flattened tracheal whole mounts stained by immunohistochemistry (Baluk et al. 2005; Yao et al. 2010) (Fig. 4.1a). Blind-ended initial lymphatics and pre-collecting lymphatics with valves are largely restricted to regions of mucosa between cartilage rings. Few lymphatics are present over cartilage rings. Large collecting lymphatics with valves and weaker LYVE-1 immunoreactivity are located downstream on the adventitial surface of the trachea. Smooth muscle cells are absent in the initial lymphatics and pre-collecting lymphatics and are sparse on collecting lymphatics. Lymphatics from the trachea join lymphatics from the esophagus and drain into mediastinal lymph nodes (Van den Broeck et al. 2006).

4.2.2 Button-Like Endothelial Junctions in Initial Lymphatics

Lymphatic endothelial cells have specialized intercellular junctions (Fig. 4.1b–d). The organization of junctional proteins is specialized to meet functional requirements, similar to those in blood vessels (Baluk et al. 2007; Dejana et al. 2009a, b). Endothelial cells of initial lymphatics are connected by discontinuous junctions called "buttons" (Fig. 4.1b). The scalloped flaps between buttons overlap each other and are thought to open and close in response to elevated interstitial fluid pressure. Live cell imaging has shown that valve-like gaps located between buttons are preferential sites of cell entry (Pflicke and Sixt 2009; Tal et al. 2011). Endothelial cells of collecting lymphatics are joined by continuous intercellular junctions called "zippers" (Baluk et al. 2007).

Understanding the nature of junctions in the initial lymphatics has evolved in multiple stages. In the 1960s, electron micrographs of thin cross sections of lymphatic vessels suggested that lymphatic endothelial cells had partially or completely open intercellular junctions (Leak and Burke 1966, 1968; Casley-Smith 1972). About ten years ago, a concept for the initial lymphatics serving as an entry or primary valve to regulate cell and fluid entry was introduced (Trzewik et al. 2001; Schmid-Schonbein 2003). Subsequent studies by immunohistochemistry with VE-cadherin and other junctional proteins in tracheal initial lymphatics demonstrated the morphological basis of primary valves (Baluk et al. 2007).

Fig. 4.1 (continued) flaps. (**e**) Development of button-like junctions from E16.5–P70 is shown as inverted gray scale images (*upper panel*, VE-cadherin) and in color (*lower panel*, VE-cadherin, *red*; LYVE-1, *green*). Scale bar: 100 μm (**a**); 20 μm (**b**, **c**); 10 μm (**e**). ((**b**, **c**, and **e**) reproduced from (Yao et al. 2012); (**d**) reproduced from (Baluk et al. 2007))

Endothelial cells of lymphatics and blood vessels are joined by two types of intercellular junctions, adherens junctions and tight junctions (Dejana et al. 2009b; Leak and Burke 1966; Komarova and Malik 2010). Buttons and zippers are composed of the same junctional proteins that include VE-cadherin at adherens junctions and occludin, claudin-5, ZO-1, JAM-A, and ESAM at tight junctions (Baluk et al. 2007).

VE-cadherin is a well-characterized marker of endothelial cell adherens junctions. VE-cadherin immunoreactivity is similarly strong in blood vessels and lymphatics (Dejana et al. 2009a, b), but the distribution of the junctional protein is different. Unlike the continuous junctions of blood vessels, VE-cadherin in initial lymphatics is distributed in the form of button-like junctions that are 3.2 μm in length and 2.9 μm apart (Baluk et al. 2007). In the trachea, buttons are most abundant in the first 750 μm from the tip of initial lymphatics and are replaced by zippers after about 1,500 μm, which is the region where pre-collecting lymphatics with valves begin.

4.2.3 Comparison of Initial Lymphatics to Collecting Lymphatics

The continuous zipper-like junctions between endothelial cells of collecting lymphatics stain for VE-cadherin at the cell border, similar to the junctions of blood vessels (Baluk et al. 2007) (Fig. 4.1c). The exclusive presence of zippers in collecting lymphatics is consistent with their importance for transport of lymph.

4.3 Plasticity of Lymphatics During Pre- and Postnatal Development

4.3.1 Dependence of Lymphatic Growth and Survival on VEGF-C/VEGFR-3 Signaling

Activation of VEGF-C/VEGFR-3 signaling is necessary for the growth and development of the lymphatic system (Lohela et al. 2009). The requirement of VEGF-C starts from the initial steps of lymphatic development when lymphatic endothelial cells sprout from venous endothelium in the early embryo (Karkkainen et al. 2004). VEGF-D is also a ligand for VEGFR-3, but appears to be dispensable in the embryo because lymphatic development proceeds normally in the absence of VEGF-D (Baldwin et al. 2005).

After proteolytic cleavage to the mature 20 kDa protein, VEGF-C can bind and activate both VEGFR-3 and VEGFR-2 (Joukov et al. 1997). Activation of VEGFR-2 can promote lymphangiogenesis (Nagy et al. 2002; Hong et al. 2004; Wirzenius et al. 2007). Activation of VEGF-C/VEGFR-3 signaling alone is usually sufficient to induce lymphangiogenesis, but VEGFR-2 and VEGFR-3 form heterodimers, and signaling through VEGFR-2/VEGFR-3 heterodimers could be involved under some conditions (Joukov et al. 1997; Nilsson et al. 2010).

Continuous signaling by VEGFR-3 is required for the survival of lymphatic endothelial cells during development, but this requirement is lost after birth (Makinen et al. 2001; Karpanen et al. 2006). Inhibition of VEGFR-3 activation can lead to regression of lymphatics until 2 weeks of postnatal age, but prolonged VEGFR-3 inhibition in the adult has no apparent effects on established lymphatics despite strong VEGFR-3 expression in lymphatic endothelial cells in the adult (Pytowski et al. 2005). The function of VEGFR-3 in maintaining the integrity of mature lymphatics deserves further investigation.

4.3.2 Transformation of Zipper- to Button-Like Junctions During the Perinatal Period

Intercellular junctions of endothelial cells can change under physiological conditions (Dejana et al. 2009a). Studies of changes in lymphatic cell junctions around the time of birth provide a better understanding of the dynamic features of lymphatics in the airways of neonatal mice (Yao et al. 2012). At E16.5, the segmental pattern of airway lymphatics is largely established, but the lymphatic endothelial cells are joined by zipper-like junctions and lack button-like junctions typical of the adult. Zippers begin to be replaced by buttons at E17.5. This transformation is particularly rapid at birth and largely complete by P28 (Fig. 4.1e). Similar perinatal transformation of zippers to buttons also occurs in the initial lymphatics of the diaphragm (Yao et al. 2012). During the transformation, the junctional proteins stay the same.

The rapid increase in number of buttons around birth is consistent with the need for efficient clearance of fluid from the lungs during the transition from an intrauterine environment of water to an external environment of air. The importance of lung lymphatics in this process is evident in transgenic mice that overexpress the extracellular domain of VEGFR-3 that traps VEGF-C and VEGF-D and thereby blocks VEGFR-3 signaling in the lungs (Kulkarni et al. 2011). These newborn mice have pulmonary lymphatic hypoplasia, increased lung weight, and high mortality.

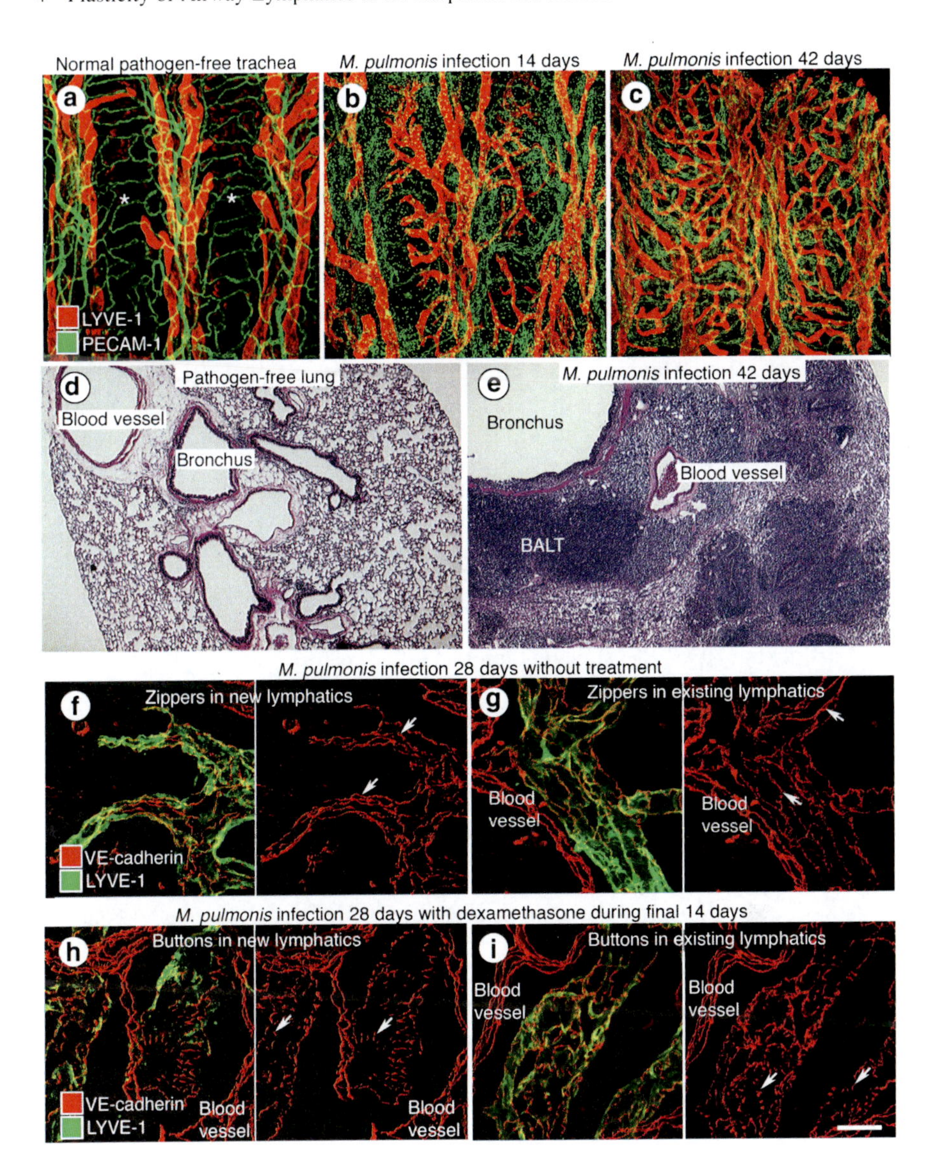

Fig. 4.2 Plasticity of lymphatics in airway inflammation. Changes in tracheal lymphatics after *M. pulmonis* infection. Confocal micrographs of mouse tracheal whole mounts stained for lymphatics (*red*, LYVE-1) and blood vessels (*green*, PECAM-1). (**a**) Few or no lymphatics are located over the cartilage rings (*asterisks*) in the pathogen-free mouse. (**b**) Tracheal lymphatics are present over cartilage rings in a mouse infected for 14 days. (**c**) Tracheal lymphatics are abundant and disorganized in a mouse infected for 42 days. H&E stained sections of mouse left lung. (**d**) No BALT is present in the pathogen-free lung. (**e**) BALT is abundant around the large bronchus and blood vessel in the lung of a mouse infected for 42 days. (**f**, **g**) Zipper-like junctions ("zippers," *arrows*) are present in the endothelium of tracheal lymphatics after infection for 28 days. (**h**, **i**) Button-like junctions ("buttons," *arrows*) are present in the endothelium of tracheal lymphatics when dexamethasone was given during the final 14 days of a 28-day infection. Scale bar: 200 μm (**a**–**c**); 400 μm (**d**, **e**); 20 μm (**f**, **i**). ((**a**–**c**) reproduced from (Yao et al. 2010); (**f**) reproduced from (Yao et al. 2012))

4.4 Plasticity of Lymphatics in Pathological Conditions

4.4.1 Lymphangiogenesis in Chronic Airway Inflammation

M. pulmonis infection has multiple attributes for studying lymphangiogenesis in sustained inflammation in the airways of mice (Lindsey and Cassell 1973). Regions overlying the cartilage rings of airways, which are normally almost free of lymphatics, have increasingly abundant lymphatics after infection (Fig. 4.2a–c). Lymphatics eventually outnumber blood vessels in the inflamed airway mucosa after *M. pulmonis* infection. Allergens have been used to sensitize and challenge the respiratory tract in studies of lung inflammation, but sustained inflammation depends on continued challenge, and few changes have been reported in respiratory lymphatics (Chu et al. 2004; Kretschmer et al. 2013).

Robust immune responses are essential for driving rapid and sustained changes in lymphatics and blood vessels of the airway mucosa after infection (Aurora et al. 2005). The lymphatic growth factors VEGF-C and VEGF-D and other cytokines released from macrophages, neutrophils, and other cells play important roles in lymphangiogenesis associated with inflammation (Baluk et al. 2005, 2009, 2013). Blocking VEGFR-3 signaling inhibits inflammatory lymphangiogenesis and reduces the enlargement of sentinel lymph nodes (Baluk et al. 2005), consistent with the contribution of cytokines, lymphatic fluid, and cell transit from inflamed airways. In this regard, defective lymphangiogenesis during airway inflammation could contribute to bronchial lymphedema and exaggerated airflow obstruction, but further studies of this issue are needed.

Bacteria or viral infection can induce immune responses that are accompanied by the development of bronchus-associated lymphoid tissue (BALT) in the lungs (Yao et al. 2010; Moyron-Quiroz et al. 2004; Rangel-Moreno et al. 2007; Kahnert et al. 2007). This so-called tertiary lymphoid tissue, consisting mainly of B-cell follicles, T cells, dendritic cells, and stromal cells, is commonly found at sites of chronic inflammation in the lung. BALT and its more peripherally located variants are abundant in the lungs of patients with chronic obstructive lung disease and are sites of lymphangiogenesis (Mori et al. 2013). BALT also accumulates in peribronchial and perivascular regions of lungs after *M. pulmonis* infection (Fig. 4.2d, e) (Yao et al. 2010). Lymphangiogenesis is reported to occur preferentially in regions of BALT (Baluk et al., unpublished findings).

4.4.2 Button-to-Zipper Transformation in Chronic Airway Inflammation

The physiological consequences of lymphangiogenesis and remodeling in chronic inflammation are poorly understood. The microvasculature of the chronically inflamed airway mucosa has abnormalities in endothelial barrier function (McDonald 1994, 2001; Schoefl 1963). The endothelium of normal blood vessels has continuous zipper-like intercellular junctions, but remodeled blood vessels have focal gaps along intercellular junctions. Remodeled blood vessels are also abnormally sensitive to inflammatory mediators that evoke plasma leakage. Mucosal edema is usually present in sustained inflammation despite widespread lymphangiogenesis. The presence of edema indicates that fluid uptake exceeds the capacity for drainage through lymphatics and other routes.

Some clues toward reconciliation of the presence of edema despite more abundant lymphatics could lie in changes in endothelial cell junctions that result in impaired fluid uptake (Baluk et al. 2005; Yao et al. 2012). In inflamed airways, newly formed lymphatics have zippers instead of buttons (Fig. 4.2f) and existing lymphatics undergo button-to-zipper transformation (Fig. 4.2g) which reverses the transformation that occurs in development (Baluk et al. 2005; Yao et al. 2012).

4.5 Reversibility of Lymphatic Growth and Remodeling

4.5.1 Reversal of Inflammation by Dexamethasone

4.5.1.1 Reversal of Lymphangiogenesis

Dexamethasone has broad anti-inflammatory activity including inhibitory effects on angiogenesis and lymphangiogenesis (Folkman and Ingber 1987; Barnes 2005). Treatment of *M. pulmonis*-infected mice with dexamethasone has distinct but different effects on blood vessels and lymphatics. Remodeled blood vessels rapidly return to their baseline state after treatment, but newly formed lymphatics are more resistant. Many new lymphatics persist after the inflammation is resolved.

After dexamethasone, most lymphatics acquire a normal, smooth contour without sprouts. Although some newly formed lymphatics undergo regression—appearing as disconnected lymphatic "islands" with little or no LYVE-1 immunoreactivity—the majority persist. Differences in the reversibility of remodeling of blood vessels and lymphatics in inflammation are also found after treatment of *M. pulmonis* infection with an antibiotic (Baluk et al. 2005).

4.5.1.2 Reversal of Button-to-Zipper Transformation

Reversal of button-to-zipper transformation is another feature of normalized lymphatics (Yao et al. 2012). Dexamethasone treatment restores the oak leaf-shaped cell phenotype typical of initial lymphatics and redistributes LYVE-1 immunoreactivity to the cell borders at sites between the segments of VE-cadherin in buttons

(Fig. 4.2h, i). This action of dexamethasone is restricted in initial lymphatics and does not change zippers in the endothelium of collecting lymphatics and thoracic duct.

The actions of dexamethasone are not limited to anti-inflammatory effects. Glucocorticoid receptor signaling has beneficial effects on perinatal lung maturation (Whitsett and Matsuzaki 2006). Glucocorticoid receptor-deficient mice develop respiratory distress and die after birth (Cole et al. 1995). The observation that dexamethasone activates phosphorylation of glucocorticoid receptors in lymphatic endothelial cells and promotes button formation in neonatal mice could represent direct actions of the steroid that complement the more generalized anti-inflammatory effects (Yao et al. 2012).

4.5.2 Reversal of Lymphangiogenesis in Other Experimental Models

The reversibility of lymphangiogenesis has not been studied as much as the growth of lymphatics, but the results are mixed. As found after *M. pulmonis* infection, newly formed lymphatics persist in skin for many months after withdrawal of overexpression of VEGF-C in transgenic mice (Lohela et al. 2008). Similarly, persistence of radiotherapy-induced lymphangiogenesis has been reported in the skin of patients with breast cancer (Jackowski et al. 2007). New lymphatics that grow in inflamed ear skin are reported to slowly regress after a wound heals (Pullinger and Florey 1937). Lymphangiogenesis in lymph nodes is similarly reported to be reversible after lymph node hypertrophy resolves (Mumprecht et al. 2012). In a suture-induced corneal inflammation model, lymphangiogenesis is described as reversible and undergoing regression more quickly than blood vessels (Cursiefen et al. 2006).

4.6 Summary and Outlook

The extraordinary plasticity of lymphatics in disease fits with the dynamic nature of airway lymphatics during perinatal development (McDonald et al. 2011). Lymphatic vessel proliferation and remodeling are also features of pulmonary lymphangiectasia, lymphangiomatosis, lymphangioleiomyomatosis (LAM), and idiopathic pulmonary fibrosis (El-Chemaly et al. 2008; Henske and McCormack 2012). Because of this plasticity, airway lymphatics serve as indicators of changing tissue requirements during normal development and in pathological conditions. Development of new animal models that recapitulate the changes would provide valuable tools for elucidating underlying mechanisms. Mechanistic insights into the driving factors and consequences of lymphatic plasticity and into the resistance of lymphatics to regression should lead to a better understanding of disease pathophysiology and new therapeutic approaches. Promotion of lymphatic growth by overexpression of VEGF-C has shown therapeutic potential by relieving the severity of skin inflammation (Huggenberger et al. 2010), improving drainage, and reducing edema in lymphedema models (Szuba et al. 2002). The same approaches could promote lymphatic maturation in other inflammation conditions. In addition, the delineation of factors that influence maturation of lymphatic endothelial cell junctions should advance the understanding of edema formation and resolution. Although it is unclear to what extent tissue edema in these conditions results from impaired lymphatic function, the use of new imaging techniques and other approaches for assessing efficiency of lymphatic fluid and cell transport should provide insights into the implications of lymphatic plasticity in disease.

Acknowledgements This work was supported in part by funding from the Lymphatic Malformation Institute, grants HL024136 and HL59157 from National Heart, Lung, and Blood Institute of the US National Institutes of Health, and the Leducq Foundation to DMcD, and by a postdoctoral fellowship award from the Lymphatic Research Foundation to LCY. We thank the members of the McDonald laboratory for critical reading of the manuscript and their helpful comments.

References

Aurora, A. B., Baluk, P., Zhang, D., Sidhu, S. S., Dolganov, G. M., Basbaum, C., et al. (2005). Immune complex-dependent remodeling of the airway vasculature in response to a chronic bacterial infection. *Journal of Immunology, 175*, 6319–6326.

Baldwin, M. E., Halford, M. M., Roufail, S., Williams, R. A., Hibbs, M. L., Grail, D., et al. (2005). Vascular endothelial growth factor D is dispensable for development of the lymphatic system. *Molecular and Cellular Biology, 25*, 2441–2449.

Baluk, P., Fuxe, J., Hashizume, H., Romano, T., Lashnits, E., Butz, S., et al. (2007). Functionally specialized junctions between endothelial cells of lymphatic vessels. *Journal of Experimental Medicine, 204*, 2349–2362.

Baluk, P., Hogmalm, A., Bry, M., Alitalo, K., Bry, K., & McDonald, D. M. (2013). Transgenic overexpression of interleukin-1beta induces persistent lymphangiogenesis but not angiogenesis in mouse airways. *American Journal of Pathology, 182*, 1434–1447.

Baluk, P., Tammela, T., Ator, E., Lyubynska, N., Achen, M. G., Hicklin, D. J., et al. (2005). Pathogenesis of persistent lymphatic vessel hyperplasia in chronic airway inflammation. *Journal of Clinical Investigation, 115*, 247–257.

Baluk, P., Yao, L. C., Feng, J., Romano, T., Jung, S. S., Schreiter, J. L., et al. (2009). TNF-alpha drives remodeling of blood vessels and lymphatics in sustained airway inflammation in mice. *Journal of Clinical Investigation, 119*, 2954–2964.

Barnes, P. J. (2005). Molecular mechanisms and cellular effects of glucocorticosteroids. *Immunology and Allergy Clinics of North America, 25*, 451–468.

Casley-Smith, J. (1972). The role of the endothelial intercellular junctions in the functioning of the intial lymphatics. *Angiologica, 9*, 106–131.

Chu, H. W., Campbell, J. A., Rino, J. G., Harbeck, R. J., & Martin, R. J. (2004). Inhaled fluticasone propionate reduces concentration of Mycoplasma pneumoniae, inflammation, and bronchial hyperresponsiveness in lungs of mice. *Journal of Infectious Diseases, 189*, 1119–1127.

Cole, T. J., Blendy, J. A., Monaghan, A. P., Krieglstein, K., Schmid, W., Aguzzi, A., et al. (1995). Targeted disruption of the glucocorticoid receptor gene blocks adrenergic chromaffin cell development and severely retards lung maturation. *Genes and Development, 9*, 1608–1621.

Cursiefen, C., Maruyama, K., Jackson, D. G., Streilein, J. W., & Kruse, F. E. (2006). Time course of angiogenesis and lymphangiogenesis after brief corneal inflammation. *Cornea, 25*, 443–447.

Dejana, E., Orsenigo, F., Molendini, C., Baluk, P., & McDonald, D. M. (2009a). Organization and signaling of endothelial cell-to-cell junctions in various regions of the blood and lymphatic vascular trees. *Cell and Tissue Research, 335*, 17–25.

Dejana, E., Tournier-Lasserve, E., & Weinstein, B. M. (2009b). The control of vascular integrity by endothelial cell junctions: molecular basis and pathological implications. *Developmental Cell, 16*, 209–221.

Dunnill, M. S. (1960). The pathology of asthma, with special reference to changes in the bronchial mucosa. *Journal of Clinical Pathology, 13*, 27–33.

Ebina, M. (2008). Remodeling of airway walls in fatal asthmatics decreases lymphatic distribution; beyond thickening of airway smooth muscle layers. *Allergology International, 57*, 165–174.

Ebina, M., Shibata, N., Ohta, H., Hisata, S., Tamada, T., Ono, M., et al. (2010). The disappearance of subpleural and interlobular lymphatics in idiopathic pulmonary fibrosis. *Lymphatic Research and Biology, 8*, 199–207.

El-Chemaly, S., Levine, S. J., & Moss, J. (2008). Lymphatics in lung disease. *Annals of the New York Academy of Sciences, 1131*, 195–202.

El-Chemaly, S., Malide, D., Zudaire, E., Ikeda, Y., Weinberg, B. A., Pacheco-Rodriguez, G., et al. (2009). Abnormal lymphangiogenesis in idiopathic pulmonary fibrosis with insights into cellular and molecular mechanisms. *Proceedings of the National Academy of Sciences of the United States of America, 106*, 3958–3963.

Folkman, J., & Ingber, D. E. (1987). Angiostatic steroids. Method of discovery and mechanism of action. *Annals of Surgery, 206*, 374–383.

Henske, E. P., & McCormack, F. X. (2012). Lymphangioleiomyomatosis: A wolf in sheep's clothing. *Journal of Clinical Investigation, 122*, 3807–3816.

Hong, Y. K., Lange-Asschenfeldt, B., Velasco, P., Hirakawa, S., Kunstfeld, R., Brown, L. F., et al. (2004). VEGF-A promotes tissue repair-associated lymphatic vessel formation via VEGFR-2 and the alpha1beta1 and alpha2beta1 integrins. *FASEB Journal, 18*, 1111–1113.

Huggenberger, R., Ullmann, S., Proulx, S. T., Pytowski, B., Alitalo, K., & Detmar, M. (2010). Stimulation of lymphangiogenesis via VEGFR-3 inhibits chronic skin inflammation. *Journal of Experimental Medicine, 207*, 2255–2269.

Jackowski, S., Janusch, M., Fiedler, E., Marsch, W. C., Ulbrich, E. J., Gaisbauer, G., et al. (2007). Radiogenic lymphangiogenesis in the skin. *American Journal of Pathology, 171*, 338–348.

Joukov, V., Sorsa, T., Kumar, V., Jeltsch, M., Claesson-Welsh, L., Cao, Y., et al. (1997). Proteolytic processing regulates receptor specificity and activity of VEGF-C. *EMBO Journal, 16*, 3898–3911.

Kahnert, A., Hopken, U. E., Stein, M., Bandermann, S., Lipp, M., & Kaufmann, S. H. (2007). Mycobacterium tuberculosis triggers formation of lymphoid structure in murine lungs. *Journal of Infectious Diseases, 195*, 46–54.

Karkkainen, M. J., Haiko, P., Sainio, K., Partanen, J., Taipale, J., Petrova, T. V., et al. (2004). Vascular endothelial growth factor C is required for sprouting of the first lymphatic vessels from embryonic veins. *Nature Immunology, 5*, 74–80.

Karpanen, T., Wirzenius, M., Makinen, T., Veikkola, T., Haisma, H. J., Achen, M. G., et al. (2006). Lymphangiogenic growth factor responsiveness is modulated by postnatal lymphatic vessel maturation. *American Journal of Pathology, 169*, 708–718.

Komarova, Y., & Malik, A. B. (2010). Regulation of endothelial permeability via paracellular and transcellular transport pathways. *Annual Review of Physiology, 72*, 463–493.

Kretschmer, S., Dethlefsen, I., Hagner-Benes, S., Marsh, L. M., Garn, H., & Konig, P. (2013). Visualization of intrapulmonary lymph vessels in healthy and inflamed murine lung using CD90/Thy-1 as a marker. *PLoS One, 8*, e55201.

Kulkarni, R. M., Herman, A., Ikegami, M., Greenberg, J. M., & Akeson, A. L. (2011). Lymphatic ontogeny and effect of hypoplasia in developing lung. *Mechanisms of Development, 128*, 29–40.
Leak, L. V., & Burke, J. F. (1966). Fine structure of the lymphatic capillary and the adjoining connective tissue area. *The American Journal of Anatomy, 118*, 785–809.
Leak, L. V., & Burke, J. F. (1968). Ultrastructural studies on the lymphatic anchoring filaments. *Journal of Cell Biology, 36*, 129–149.
Lindsey, J. R., & Cassell, H. (1973). Experimental Mycoplasma pulmonis infection in pathogen-free mice. Models for studying mycoplasmosis of the respiratory tract. *American Journal of Pathology, 72*, 63–90.
Lohela, M., Bry, M., Tammela, T., & Alitalo, K. (2009). VEGFs and receptors involved in angiogenesis versus lymphangiogenesis. *Current Opinion in Cell Biology, 21*, 154–165.
Lohela, M., Helotera, H., Haiko, P., Dumont, D. J., & Alitalo, K. (2008). Transgenic induction of vascular endothelial growth factor-C is strongly angiogenic in mouse embryos but leads to persistent lymphatic hyperplasia in adult tissues. *American Journal of Pathology, 173*, 1891–1901.
Makinen, T., Jussila, L., Veikkola, T., Karpanen, T., Kettunen, M. I., Pulkkanen, K. J., et al. (2001). Inhibition of lymphangiogenesis with resulting lymphedema in transgenic mice expressing soluble VEGF receptor-3. *Nature Medicine, 7*, 199–205.
Mandal, R. V., Mark, E. J., & Kradin, R. L. (2008). Organizing pneumonia and pulmonary lymphatic architecture in diffuse alveolar damage. *Human Pathology, 39*, 1234–1238.
McDonald, D. M. (1994). Endothelial gaps and permeability of venules in rat tracheas exposed to inflammatory stimuli. *American Journal of Physiology, 266*, L61–L83.
McDonald, D. M. (2001). Angiogenesis and remodeling of airway vasculature in chronic inflammation. *American Journal of Respiratory and Critical Care Medicine, 164*, S39–S45.
McDonald, D. M. (2008). Angiogenesis and vascular remodeling in inflammation and cancer: biology and architecture of the vasculature. In W. D. Figg & J. Folkman (Eds.), *Angiogenesis: an integrative approach from science to medicine* (pp. 17–33). New York: Springer. Chapter 2.
McDonald, D. M., Yao, L. C., & Baluk, P. (2011). Dynamics of airway blood vessels and lymphatics: Lessons from development and inflammation. *Proceedings of the American Thoracic Society, 8*, 504–507.
Mori, M., Andersson, C. K., Svedberg, K. A., Glader, P., Bergqvist, A., Shikhagaie, M., et al. (2013). Appearance of remodelled and dendritic cell-rich alveolar-lymphoid interfaces provides a structural basis for increased alveolar antigen uptake in chronic obstructive pulmonary disease. *Thorax, 68*(6), 521–531.
Moyron-Quiroz, J. E., Rangel-Moreno, J., Kusser, K., Hartson, L., Sprague, F., Goodrich, S., et al. (2004). Role of inducible bronchus associated lymphoid tissue (iBALT) in respiratory immunity. *Nature Medicine, 10*, 927–934.
Mumprecht, V., Roudnicky, F., & Detmar, M. (2012). Inflammation-induced lymph node lymphangiogenesis is reversible. *American Journal of Pathology, 180*, 874–879.
Nagy, J. A., Vasile, E., Feng, D., Sundberg, C., Brown, L. F., Detmar, M. J., et al. (2002). Vascular permeability factor/vascular endothelial growth factor induces lymphangiogenesis as well as angiogenesis. *Journal of Experimental Medicine, 196*, 1497–1506.
Nilsson, I., Bahram, F., Li, X., Gualandi, L., Koch, S., Jarvius, M., et al. (2010). VEGF receptor 2/-3 heterodimers detected in situ by proximity ligation on angiogenic sprouts. *EMBO Journal, 29*, 1377–1388.
Parra, E. R., Araujo, C. A., Lombardi, J. G., Ab'Saber, A. M., Carvalho, C. R., Kairalla, R. A., & Capelozzi, V. L. (2012). Lymphatic fluctuation in the parenchymal remodeling stage of acute interstitial pneumonia, organizing pneumonia, nonspecific interstitial pneumonia and idiopathic pulmonary fibrosis. *Brazilian Journal of Medical and Biological Research, 45*, 466–472.
Pflicke, H., & Sixt, M. (2009). Preformed portals facilitate dendritic cell entry into afferent lymphatic vessels. *Journal of Experimental Medicine, 206*, 2925–2935.

Pullinger, B., & Florey, H. W. (1937). Proliferation of lymphatics in inflammation. *Journal of Pathology and Bacteriology, 45*, 157–170.

Pytowski, B., Goldman, J., Persaud, K., Wu, Y., Witte, L., Hicklin, D. J., et al. (2005). Complete and specific inhibition of adult lymphatic regeneration by a novel VEGFR-3 neutralizing antibody. *Journal of the National Cancer Institute, 97*, 14–21.

Rangel-Moreno, J., Moyron-Quiroz, J. E., Hartson, L., Kusser, K., & Randall, T. D. (2007). Pulmonary expression of CXC chemokine ligand 13, CC chemokine ligand 19, and CC chemokine ligand 21 is essential for local immunity to influenza. *Proceedings of the National Academy of Sciences of the United States of America, 104*, 10577–10582.

Schmid-Schonbein, G. W. (2003). The second valve system in lymphatics. *Lymphatic Research and Biology, 1*, 25–29. discussion 29–31.

Schoefl, G. I. (1963). Studies on inflammation. III. Growing capillaries: their structure and permeability. *Virchows Archiv für Pathologische Anatomie und Physiologie und für Klinische Medizin, 337*, 97–141.

Szuba, A., Skobe, M., Karkkainen, M. J., Shin, W. S., Beynet, D. P., Rockson, N. B., et al. (2002). Therapeutic lymphangiogenesis with human recombinant VEGF-C. *FASEB Journal, 16*, 1985–1987.

Tal, O., Lim, H. Y., Gurevich, I., Milo, I., Shipony, Z., Ng, L. G., et al. (2011). DC mobilization from the skin requires docking to immobilized CCL21 on lymphatic endothelium and intralymphatic crawling. *Journal of Experimental Medicine, 208*, 2141–2153.

Trzewik, J., Mallipattu, S. K., Artmann, G. M., Delano, F. A., & Schmid-Schonbein, G. W. (2001). Evidence for a second valve system in lymphatics: Endothelial microvalves. *FASEB Journal, 15*, 1711–1717.

Van den Broeck, W., Derore, A., & Simoens, P. (2006). Anatomy and nomenclature of murine lymph nodes: Descriptive study and nomenclatory standardization in BALB/cAnNCrl mice. *Journal of Immunological Methods, 312*, 12–19.

Whitsett, J. A., & Matsuzaki, Y. (2006). Transcriptional regulation of perinatal lung maturation. *Pediatric Clinics of North America, 53*, 873–887. viii.

Wilson, J. W., & Hii, S. (2006). The importance of the airway microvasculature in asthma. *Current Opinion in Allergy and Clinical Immunology, 6*, 51–55.

Wirzenius, M., Tammela, T., Uutela, M., He, Y., Odorisio, T., Zambruno, G., et al. (2007). Distinct vascular endothelial growth factor signals for lymphatic vessel enlargement and sprouting. *Journal of Experimental Medicine, 204*, 1431–1440.

Yamashita, M., Iwama, N., Date, F., Chiba, R., Ebina, M., Miki, H., et al. (2009). Characterization of lymphangiogenesis in various stages of idiopathic diffuse alveolar damage. *Human Pathology, 40*, 542–551.

Yao, L. C., Baluk, P., Feng, J., & McDonald, D. M. (2010). Steroid-resistant lymphatic remodeling in chronically inflamed mouse airways. *American Journal of Pathology, 176*, 1525–1541.

Yao, L. C., Baluk, P., Srinivasan, R. S., Oliver, G., & McDonald, D. M. (2012). Plasticity of button-like junctions in the endothelium of airway lymphatics in development and inflammation. *American Journal of Pathology, 180*, 2561–2575.

Chapter 5
Regulation of Lymphatic Vasculature by Extracellular Matrix

Sophie Lutter and Taija Makinen

Abstract The extracellular matrix (ECM) is a complex but highly organized network of macromolecules with different physical, biochemical, and mechanical properties. In addition to providing structural support to tissues, it regulates a variety of cellular responses during development and tissue homeostasis. Interactions between the lymphatic vessels and their ECM are starting to be recognized as important modulators of lymphangiogenesis. Here, we review the current knowledge of the structure and composition of the ECM of lymphatic vessels and discuss the role of individual matrix components and their cell surface receptors in regulating lymphatic vascular development and function.

5.1 The Extracellular Matrix and Vascular Basement Membrane

The extracellular matrix (ECM), a complex meshwork of proteins and sugars, provides a protective and structural scaffold for cells, tissues, and organs. The major structural components of the ECM are collagens and glycosaminoglycans, which include hyaluronan and different types of proteoglycans (PGs) that are further classified based on their core proteins, distribution, or glycosaminoglycan content (heparan sulfate PG, chondroitin sulfate PG, or dermatan sulfate PG) (Raman et al. 2005; Wiig et al. 2010). An example of specialized ECM is found in the vascular basement membrane (BM), which is an acellular, sheetlike structure, associated with the cells of the vessel wall. Vascular basement membranes, which

S. Lutter
Lymphatic Development Laboratory, Cancer Research UK London Research Institute, 44 Lincoln's Inn Fields, London WC2A 3LY, UK

T. Makinen (✉)
Uppsala University, Dept. Immunology, Genetics and Pathology, Dag Hammarskjöldsv. 20 751 85, Uppsala, Sweden
e-mail: taija.makinen@igp.uu.se

F. Kiefer and S. Schulte-Merker (eds.), *Developmental Aspects of the Lymphatic Vascular System*, Advances in Anatomy, Embryology and Cell Biology 214, DOI 10.1007/978-3-7091-1646-3_5,

are produced and shared by endothelial cells (ECs) and associated mural cells, consist predominantly of collagen IV, fibronectin (FN), nidogen/entactin, laminins, and heparan sulfate PGs (HSPGs).

Mouse knockout models have demonstrated the importance and distinct functions of several individual BM components (Aszodi et al. 2006). For example, loss of fibronectin results in loss or severe deformity of blood vessels and defective embryonic vasculature (Astrof and Hynes 2009). In contrast, collagen IV is dispensable for early vascular morphogenesis and the knockout embryos show rather mild vascular phenotype with irregular protrusion of capillaries into the neural layers (Poschl et al. 2004). Like certain other BMs (Poschl et al. 2004), vascular BMs may require collagen IV for the maintenance of their integrity. Complete lack of extracellular laminins was still compatible with the formation of vessels and their organization into three-dimensional structures with pericyte coverage in spheroids of differentiating embryonic stem cells (Jakobsson et al. 2008). However, laminin-deficient vessels showed widening of the vascular lumen as well as changes in BM composition, which may compromise the stability of the vessels under in vivo conditions where flow and mechanical forces are present (Jakobsson et al. 2008).

Apart from conferring structural support and stability to blood vessels, the ECM also functions to regulate cell behavior such as migration, differentiation, polarity, survival, and proliferation (Kruegel and Miosge 2010). It can mediate efficient communication between endothelial cells, or endothelial and smooth muscle cells (SMCs), without the need for direct cell–cell contacts (Kruegel and Miosge 2010). These functions are mediated via binding of cells to the matrix via specific cell surface receptors, including integrins (binding to the major ECM ligands: fibronectin, vitronectin, collagens, laminins), discoidin domain receptors (binding to collagens), or cell surface PGs (binding to, e.g., fibronectin) (Hynes 2007; Wiig et al. 2010). In addition, by binding to growth factors, ECM components can either positively or negatively regulate cell behavior (Hynes 2009). For example, HSPGs bind to the pro-angiogenic growth factor VEGF, thus providing spatially regulated angiogenic cues to direct vascular morphogenesis (Gerhardt et al. 2003; Ruhrberg et al. 2002). The requirement for the heparan sulfate-binding domain on PDGF-BB for pericyte recruitment and PDGF-BB retention at the cell surface demonstrates another example of the ECM positively regulating growth factor signaling (Abramsson et al. 2007; Abramsson et al. 2003). The ECM can also negatively regulate cell behavior, as in the case of HSPG sequestration of HB-EGF, which promotes juxtacrine HB-EGF signaling leading to growth inhibition (Prince et al. 2010).

5.2 Extracellular Matrix of Lymphatic Vessels

In contrast to the blood vasculature, much less is known about the composition and function of the lymphatic BM. It was long believed that lymphatic vessels were devoid of BMs, but recent studies have shown their presence in all calibers of vessels (Lutter et al. 2012; Norrmen et al. 2009; Pflicke and Sixt 2009; Vainionpaa et al. 2007). The larger lymphatic vessels, called collecting vessels, have a continuous BM composed of collagen IV, fibronectin, and laminins ((Lutter et al. 2012; Norrmen et al. 2009), Figs. 5.1 and 5.2a–c). Recruitment of SMCs to the collecting vessels coincides with the assembly of BMs (Lutter et al. 2012), similar to what has been reported during blood vessel development (Stratman et al. 2009). In vitro experiments further showed that the amount of BM proteins produced by both blood endothelial cells and SMCs is increased upon cell–cell contact (Stratman et al. 2009). This suggests an important role of SMCs in blood vessel wall assembly that is likely to apply to lymphatic vessels as well. In addition to the BM of the vessel wall, luminal valves present in the collecting lymphatic vessels contain an ECM core on which the valve endothelial cells attach (Fig. 5.1). This specialized structure has a unique composition different from that of the BM of the vessel wall, containing high levels of laminin-α5 and fibronectin-EIIIA/EDA (*FN-EIIIA*) splice isoform ((Bazigou et al. 2009), Fig 5.2a, b).

The smaller lymphatic capillaries (also called initial lymphatic vessels) lack SMCs and luminal valves. The composition of the BM of lymphatic capillaries is therefore likely to differ from that of collecting lymphatic vessels where both ECs and SMCs contribute to the production of the ECM proteins. However, there has been little investigation of the difference between BMs of the two types of vessels and how this might influence their individual vessel functions. Existing data show that lymphatic capillaries have a discontinuous BM containing gaps that allow entry of fluid and immune cells into the vessel lumen ((Pflicke and Sixt 2009), Figs. 5.1 and 5.2d). To further facilitate fluid entry under conditions of high interstitial pressure, the capillary lymphatic endothelial cells (LECs) are attached to the interstitial matrix by anchoring filaments composed of microfibril bundles and elastin that prevent the vessels from collapsing ((Paupert et al. 2011), Fig. 5.1, see also Chap. 3 of this issue).

5.3 Regulation of Lymphangiogenesis by Cell–Matrix Interactions

5.3.1 Lymphatic Vessel Sprouting and Growth

Extracellular matrix can directly guide vascular growth by providing adhesive gradients for directional migration of cells. In addition, gradients of chemoattractant molecules present in the matrix provide interstitial guidance for migrating cells. A well-described example of the latter is the abovementioned ECM retention of the heparin-binding isoforms of the major angiogenic growth factor VEGF, which is critical for the correct guidance of angiogenic vessel sprouts and patterning of the vasculature (Gerhardt et al. 2003; Ruhrberg et al. 2002). Two other VEGF family members, VEGF-C and VEGF-D, regulate lymphatic vessel sprouting. In particular, VEGF-C is required for the sprouting of lymphatic vessels

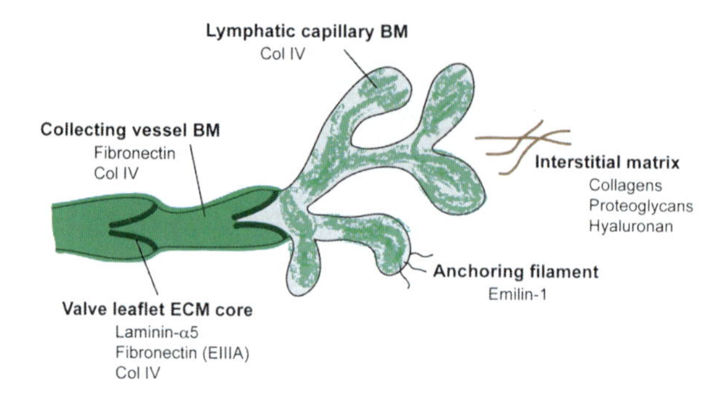

Fig. 5.1 Schematic view of the extracellular matrix of lymphatic vessels. Extracellular matrices around lymphatic vessels and their key molecular components are shown. Collecting vessels have a continuous basement membrane (BM) while that of lymphatic capillaries contains gaps and discontinuities, which allow contact and anchoring of LECs to the interstitial matrix. Luminal valve leaflets contain a specialized ECM core on which valve endothelial cells are attached

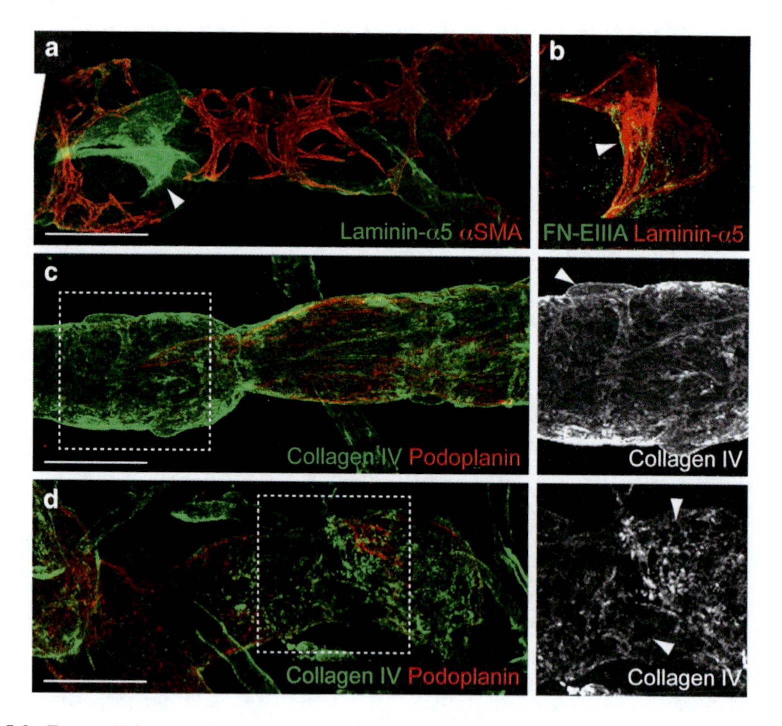

Fig. 5.2 Extracellular matrix components of the lymphatic basement membrane. Whole-mount immunofluorescence of collecting lymphatic vessels (**a–c**) and lymphatic capillaries (**d**) for indicated antibodies to ECM proteins or markers of SMC (αSMA) or LEC (Podoplanin). Single-channel images show collagen IV staining of the *boxed regions* shown in (**c**, **d**). *Arrowheads* in (**a**, **b**) point to luminal valves, in (**c**) to collagen IV-positive SMC, and in (**d**) to gaps in the BM of lymphatic capillaries. Scale bars: 50 μm

from the veins during early embryonic development (Hagerling et al. 2013; Karkkainen et al. 2004), and it also promotes lymphatic vessel growth in postnatal development and in pathological conditions (reviewed in Alitalo 2011). In contrast to VEGF, VEGF-C and VEGF-D do not have a heparin-binding domain, but they instead contain a carboxy-terminal silk homology domain that requires proteolytic cleavage for full activation. VEGF-C was nevertheless shown to bind to heparan sulfate, which was required for full VEGF-C activity on cultured LECs (Yin et al. 2011). However, it is not known if VEGF-C gradients are present in tissues in vivo and required for the guidance of lymphangiogenic vessel sprouts. If gradients of VEGF-C are indeed present, the mechanism of their generation and their shape are likely different from that of VEGF. This is suggested by a finding that a chimeric growth factor consisting of the receptor-binding domain of VEGF-C fused to the heparin-binding domain of VEGF that specifically targeted lymphatic vessel growth to heparan sulfate-rich basement membranes induces a different architecture of lymphatic vessels compared to that caused by native VEGF-C (Tammela et al. 2007).

Collagen and calcium-binding EGF domain-1 (*CCBE1*) is another key regulator of early lymphatic sprouting from the veins (Bos et al. 2011; Hogan et al. 2009). It is a secreted protein, produced by tissues in close proximity to the sprouting LECs, which is predicted to bind components of the ECM. However, its receptor and binding partners are yet to be identified. In mice *Ccbe1* deficiency leads to aberrant sprout formation, which is followed by downregulation of the homeobox transcription factor *Prox1*, a master regulator of LEC identity, followed by rapid loss of all lymphatic structures ((Hagerling et al. 2013) also see Chap. 13 of this issue). Although *CCBE1* does not appear to have lymphangiogenic activity on its own, it enhances the effect of VEGF-C in vivo (Bos et al. 2011).

Not only the structural components of the matrix but also the extracellular fluid within the interstitial space can regulate lymphangiogenic processes. It was recently demonstrated that the increase in the volume of interstitial fluid provides an important signal for the expansion of the lymphatic vasculature during early development (Planas-Paz et al. 2012). Increased fluid pressure was shown to induce mechanical stretching of cells that led to integrin-β1-dependent activation of VEGFR-3 signaling and LEC proliferation ((Planas-Paz et al. 2012) and Chap. 3 of this issue).

5.3.2 *Collecting Vessel Formation and Function*

Several ECM components and their receptors have been implicated in the remodeling and maturation of the lymphatic vasculature as it acquires a hierarchical organization of vessels with distinct functions. For example, integrin-α9 and its ECM ligand, fibronectin containing the alternatively spliced EIIIA/EDA domain, play an important role in collecting lymphatic vessel formation and valve morphogenesis. Integrin-α9 is highly expressed in the developing and mature lymphatic valves (Bazigou et al. 2009). *Itgα9* (encoding integrin-α9)-deficient mice develop only rudimentary valve structures that show a disorganized ECM core within the valve leaflets and are incapable of preventing lymph backflow (Bazigou et al. 2009). Defective valve formation in *FN-EIIIA* mutant mice and disorganization of the *FN-EIIIA* matrix in *Itgα9*-deficient mice further identified *FN-EIIIA* as the physiologically relevant integrin-α9 ligand in the development of lymphatic valves (Bazigou et al. 2009). More recent work has indicated a function for tumor-derived *FN-EIIIA* in enhancing LEC tubulogenesis and branching in vitro, suggesting other potential roles for this ECM protein in tumor-induced lymphangiogenesis (Ou et al. 2010). Other matrix components of the valve leaflet include laminin-α5, which is deposited at the earliest stages of valve formation (Bazigou et al. 2009; Sabine et al. 2012). However, the function of laminin-α5 in valve morphogenesis is not known.

Collecting lymphatic vessel maturation is also associated with the deposition of a continuous layer of BM around these vessels. While many of the major components are shared between blood and lymphatic vessels, several ECM components that are specific to or enriched in lymphatic in comparison to vascular BM were recently identified (Lutter et al. 2012). One of these components is Reelin, a large secreted glycoprotein that has been previously implicated as a critical regulator of neuronal migration and patterning of the cortex (reviewed in Forster et al. 2010). In the lymphatic vasculature Reelin is expressed in LECs of all vessel types, but it is efficiently secreted to the ECM of collecting vessels only. Reelin-deficiency resulted in reduced SMC coverage and abnormal collecting vessel morphology and function, suggesting its critical role in the formation of functional lymphatic vasculature (Lutter et al. 2012).

5.3.3 Lymphatic Capillary Formation and Function

As discussed above, due to the complete lack of mural cells in lymphatic capillaries, the composition and organization of their BMs are likely different from all other vascular BMs, including collecting lymphatic vessels and blood capillaries that have either sparse SMC or pericyte coverage, respectively. Expression analyses show the presence of several laminins, collagen IV and XVIII, and nidogen-1 around lymphatic capillaries (Lutter et al. 2012; Vainionpaa et al. 2007). Discontinuities and gaps in the BM of lymphatic capillaries allow passage of immune cells (Baluk et al. 2007) and are likely important for the maintenance of permeability of lymphatic capillaries to fluid and macromolecules. In agreement, increased deposition of BM components in lymphatic capillaries, which is observed in connection with ectopic SMC recruitment in several mouse mutants, can lead to abnormal and/or nonfunctional lymphatic vessels (Dellinger et al. 2008; Makinen et al. 2005; Petrova et al. 2004; Saharinen et al. 2010). Defects in the ECM anchorage of capillary LECs have also been shown to result in reduced lymphatic function. Loss of Emilin-1, a major component of the anchoring filaments attaching lymphatic capillaries to the interstitial matrix, leads to hyperplastic lymphatic capillaries that fail to respond to changes in interstitial pressure (Danussi et al. 2008).

LECs of the lymphatic capillaries specifically express high levels of Lyve1, a receptor for hyaluronan, which is the major ECM glycosaminoglycan (Banerji et al. 1999). However, the functional significance of this interaction remains unclear, mainly due to the lack of an obvious phenotype in *Lyve1*-deficient mice (Gale et al. 2007).

5.3.4 Pathological Lymphangiogenesis

The ECM is often deregulated in pathological conditions and can promote the progression of several diseases, including cancer. Several integrins have been implicated in pathological lymphangiogenesis. For example, small molecule inhibitors for integrin-α5, a major FN receptor, reduced lymphangiogenesis in vivo in a model of inflammatory corneal lymphangiogenesis. Interestingly, blood endothelial cells were less sensitive to integrin-α5 inhibition and at small inhibitor doses lymphangiogenesis could be selectively inhibited (Dietrich et al. 2007). Integrin-α4β1, another FN receptor, has also been implicated in both growth factor and tumor-induced lymphangiogenesis. It is expressed only on proliferative LECs and lymphatic vessels and promotes LEC adhesion and migration on cellular fibronectin in vitro. In vivo, VEGF-C-induced lymphangiogenesis was inhibited by integrin-α 4β1 antagonists or in mice lacking integrin-α4 (Garmy-Susini et al. 2010). In addition, the collagen and laminin receptors integrin-α1β2 and integrin-α2β1 were shown to participate in VEGF-A-mediated lymphangiogenesis in wound healing (Hong et al. 2004). Collagen I was further shown to induce lymphatic regeneration after skin wounding via downregulation of transforming growth factor-β1 (TGFβ1) expression and increased LEC migration and proliferation (Clavin et al. 2008). In contrast, anti-lymphangiogenic properties have been suggested for certain other collagens, such as collagen XVIII and its proteolytic cleavage product, neostatin-7 (Kojima et al. 2008). Endostatin, a proteolytic fragment of collagen XVIII, was shown to inhibit tumor lymphangiogenesis, but via an indirect mechanism mediated by inhibition of mast cell migration and adhesion and the resulting decrease in tumoral VEGF-C levels (Brideau et al. 2007).

5.4 Matrix Remodeling and Lymphangiogenesis

Matrix metalloproteinases (MMPs) play important roles in regulating cellular responses to the ECM by processing individual BM proteins to release soluble or active fragments. In addition, MMP-mediated degradation of the ECM is an important first step in the induction of angiogenesis, allowing the endothelial cells greater mobility to sprout from the existing vessel (Arroyo and Iruela-Arispe 2010). The role of MMPs in lymphangiogenesis is however yet to be clearly defined. Genetic profiling of cultured endothelial cells has shown that LECs produce fewer and lower levels of proteases, such as MMPs, compared to blood ECs (Petrova et al. 2002; Podgrabinska et al. 2002). Two MMPs, MMP-9 and MMP-2, were found to be strongly upregulated in the vicinity of lymphatic regeneration in a skin wound model in vivo. However, the expression of MMP-9 declined before, while MMP-2 levels peaked at the onset of lymphangiogenesis. Normal lymphatic regeneration in *Mmp9*$^{-/-}$ mice further suggested that this protease is not involved in lymphangiogenesis (Rutkowski et al. 2006). In contrast, a role for MMP-2 in lymphangiogenesis was supported by reduced lymphatic vessel sprouting in vitro and ex vivo by inhibition of MT1-MMP-catalyzed activation of proMMP-2 (Ingvarsen et al. 2013).

5.5 Concluding Remarks

Although the importance of ECM proteins in lymphatic development and function is starting to be recognized, there is still much more work to do to fully understand the composition and function of the lymphatic basement membrane, how this may differ between lymphatic capillaries and collecting vessels, and how the basement membrane can affect lymphatic vessel development and function.

Acknowledgements We thank Lars Jakobsson, Sally Leevers, and Ingvar Ferby for comments on the manuscript. The work in our laboratory is supported by Cancer Research UK and EMBO Young Investigator Programme.

References

Abramsson, A., Kurup, S., Busse, M., Yamada, S., Lindblom, P., Schallmeiner, E., et al. (2007). Defective N-sulfation of heparan sulfate proteoglycans limits PDGF-BB binding and pericyte recruitment in vascular development. *Genes and Development, 21*(3), 316–331.

Abramsson, A., Lindblom, P., & Betsholtz, C. (2003). Endothelial and nonendothelial sources of PDGF-B regulate pericyte recruitment and influence vascular pattern formation in tumors. *Journal of Clinical Investigation, 112*(8), 1142–1151.

Alitalo, K. (2011). The lymphatic vasculature in disease. *Nature Medicine, 17*(11), 1371–1380.

Arroyo, A. G., & Iruela-Arispe, M. L. (2010). Extracellular matrix, inflammation, and the angiogenic response. *Cardiovascular Research, 86*(2), 226–235.

Astrof, S., & Hynes, R. O. (2009). Fibronectins in vascular morphogenesis. *Angiogenesis, 12*(2), 165–175.

Aszodi, A., Legate, K. R., Nakchbandi, I., & Fassler, R. (2006). What mouse mutants teach us about extracellular matrix function. *Annual Review of Cell and Developmental Biology, 22*, 591–621.

Baluk, P., Fuxe, J., Hashizume, H., Romano, T., Lashnits, E., Butz, S., et al. (2007). Functionally specialized junctions between endothelial cells of lymphatic vessels. *Journal of Experimental Medicine, 204*(10), 2349–2362.

Banerji, S., Ni, J., Wang, S. X., Clasper, S., Su, J., Tammi, R., et al. (1999). LYVE-1, a new homologue of the CD44 glycoprotein, is a lymph-specific receptor for hyaluronan. *Journal of Cell Biology, 144*(4), 789–801.

Bazigou, E., Xie, S., Chen, C., Weston, A., Miura, N., Sorokin, L., et al. (2009). Integrin-alpha9 is required for fibronectin matrix assembly during lymphatic valve morphogenesis. *Developmental Cell, 17*(2), 175–186.

Bos, F. L., Caunt, M., Peterson-Maduro, J., Planas-Paz, L., Kowalski, J., Karpanen, T., et al. (2011). CCBE1 is essential for mammalian lymphatic vascular development and enhances the lymphangiogenic effect of vascular endothelial growth factor-C in vivo. *Circulation Research, 109*(5), 486–491.

Brideau, G., Makinen, M. J., Elamaa, H., Tu, H., Nilsson, G., Alitalo, K., et al. (2007). Endostatin overexpression inhibits lymphangiogenesis and lymph node metastasis in mice. *Cancer Research, 67*(24), 11528–11535.

Clavin, N. W., Avraham, T., Fernandez, J., Daluvoy, S. V., Soares, M. A., Chaudhry, A., & Mehrara, B. J. (2008). TGF-beta1 is a negative regulator of lymphatic regeneration during wound repair. *American Journal of Physiology. Heart and Circulatory Physiology, 295*(5), H2113–H2127.

Danussi, C., Spessotto, P., Petrucco, A., Wassermann, B., Sabatelli, P., Montesi, M., et al. (2008). Emilin1 deficiency causes structural and functional defects of lymphatic vasculature. *Molecular and Cellular Biology, 28*(12), 4026–4039.

Dellinger, M., Hunter, R., Bernas, M., Gale, N., Yancopoulos, G., Erickson, R., & Witte, M. (2008). Defective remodeling and maturation of the lymphatic vasculature in Angiopoietin-2 deficient mice. *Developmental Biology, 319*(2), 309–320.

Dietrich, T., Onderka, J., Bock, F., Kruse, F. E., Vossmeyer, D., Stragies, R., et al. (2007). Inhibition of inflammatory lymphangiogenesis by integrin alpha5 blockade. *American Journal of Pathology, 171*(1), 361–372.

Forster, E., Bock, H. H., Herz, J., Chai, X., Frotscher, M., & Zhao, S. (2010). Emerging topics in Reelin function. *European Journal of Neuroscience, 31*(9), 1511–1518.

Gale, N. W., Prevo, R., Espinosa, J., Ferguson, D. J., Dominguez, M. G., Yancopoulos, G. D., et al. (2007). Normal lymphatic development and function in mice deficient for the lymphatic hyaluronan receptor LYVE-1. *Molecular and Cellular Biology, 27*(2), 595–604.

Garmy-Susini, B., Avraamides, C. J., Schmid, M. C., Foubert, P., Ellies, L. G., Barnes, L., et al. (2010). Integrin alpha4beta1 signaling is required for lymphangiogenesis and tumor metastasis. *Cancer Research, 70*(8), 3042–3051.

Gerhardt, H., Golding, M., Fruttiger, M., Ruhrberg, C., Lundkvist, A., Abramsson, A., et al. (2003). VEGF guides angiogenic sprouting utilizing endothelial tip cell filopodia. *Journal of Cell Biology, 161*(6), 1163–1177.

Hagerling, R., Pollmann, C., Andreas, M., Schmidt, C., Nurmi, H., Adams, R. H., et al. (2013). A novel multistep mechanism for initial lymphangiogenesis in mouse embryos based on ultramicroscopy. *EMBO Journal, 32*(5), 629–644.

Hogan, B. M., Bos, F. L., Bussmann, J., Witte, M., Chi, N. C., Duckers, H. J., & Schulte-Merker, S. (2009). CCBE1 is required for embryonic lymphangiogenesis and venous sprouting. *Nature Genetics, 41*(4), 396–398.

Hong, Y. K., Lange-Asschenfeldt, B., Velasco, P., Hirakawa, S., Kunstfeld, R., Brown, L. F., et al. (2004). VEGF-A promotes tissue repair-associated lymphatic vessel formation via VEGFR-2 and the alpha1beta1 and alpha2beta1 integrins. *FASEB Journal, 18*(10), 1111–1113.

Hynes, R. O. (2007). Cell-matrix adhesion in vascular development. *Journal of Thrombosis and Haemostasis, 5*(Suppl 1), 32–40.

Hynes, R. O. (2009). The extracellular matrix: Not just pretty fibrils. *Science, 326*(5957), 1216–1219.

Ingvarsen, S., Porse, A., Erpicum, C., Maertens, L., Jurgensen, H. J., Madsen, D. H., et al. (2013). Targeting a single function of the multifunctional matrix metalloprotease MT1-MMP. Impact on lymphangiogenesis. *Journal of Biological Chemistry, 288*(15), 10195–10204.

Jakobsson, L., Domogatskaya, A., Tryggvason, K., Edgar, D., & Claesson-Welsh, L. (2008). Laminin deposition is dispensable for vasculogenesis but regulates blood vessel diameter independent of flow. *FASEB Journal, 22*(5), 1530–1539.

Karkkainen, M. J., Haiko, P., Sainio, K., Partanen, J., Taipale, J., Petrova, T. V., et al. (2004). Vascular endothelial growth factor C is required for sprouting of the first lymphatic vessels from embryonic veins. *Nature Immunology, 5*(1), 74–80.

Kojima, T., Azar, D. T., & Chang, J. H. (2008). Neostatin-7 regulates bFGF-induced corneal lymphangiogenesis. *FEBS Letters, 582*(17), 2515–2520.

Kruegel, J., & Miosge, N. (2010). Basement membrane components are key players in specialized extracellular matrices. *Cellular and Molecular Life Sciences, 67*(17), 2879–2895.

Lutter, S., Xie, S., Tatin, F., & Makinen, T. (2012). Smooth muscle-endothelial cell communication activates Reelin signaling and regulates lymphatic vessel formation. *Journal of Cell Biology, 197*(6), 837–849.

Makinen, T., Adams, R. H., Bailey, J., Lu, Q., Ziemiecki, A., Alitalo, K., et al. (2005). PDZ interaction site in ephrinB2 is required for the remodeling of lymphatic vasculature. *Genes and Development, 19*(3), 397–410.

Norrmen, C., Ivanov, K. I., Cheng, J., Zangger, N., Delorenzi, M., Jaquet, M., et al. (2009). FOXC2 controls formation and maturation of lymphatic collecting vessels through cooperation with NFATc1. *Journal of Cell Biology, 185*(3), 439–457.

Ou, J. J., Wu, F., & Liang, H. J. (2010). Colorectal tumor derived fibronectin alternatively spliced EDA domain exserts lymphangiogenic effect on human lymphatic endothelial cells. *Cancer Biology and Therapy, 9*(3), 186–191.

Paupert, J., Sounni, N. E., & Noel, A. (2011). Lymphangiogenesis in post-natal tissue remodeling: lymphatic endothelial cell connection with its environment. *Molecular Aspects of Medicine, 32*(2), 146–158.

Petrova, T. V., Karpanen, T., Norrmen, C., Mellor, R., Tamakoshi, T., Finegold, D., et al. (2004). Defective valves and abnormal mural cell recruitment underlie lymphatic vascular failure in lymphedema distichiasis. *Nature Medicine, 10*(9), 974–981.

Petrova, T. V., Makinen, T., Makela, T. P., Saarela, J., Virtanen, I., Ferrell, R. E., et al. (2002). Lymphatic endothelial reprogramming of vascular endothelial cells by the Prox-1 homeobox transcription factor. *EMBO Journal, 21*(17), 4593–4599.

Pflicke, H., & Sixt, M. (2009). Preformed portals facilitate dendritic cell entry into afferent lymphatic vessels. *Journal of Experimental Medicine, 206*(13), 2925–2935.

Planas-Paz, L., Strilic, B., Goedecke, A., Breier, G., Fassler, R., & Lammert, E. (2012). Mechanoinduction of lymph vessel expansion. *EMBO Journal, 31*(4), 788–804.

Podgrabinska, S., Braun, P., Velasco, P., Kloos, B., Pepper, M. S., & Skobe, M. (2002). Molecular characterization of lymphatic endothelial cells. *Proceedings of the National Academy of Sciences of the United States of America, 99*(25), 16069–16074.

Poschl, E., Schlotzer-Schrehardt, U., Brachvogel, B., Saito, K., Ninomiya, Y., & Mayer, U. (2004). Collagen IV is essential for basement membrane stability but dispensable for initiation of its assembly during early development. *Development, 131*(7), 1619–1628.

Prince, R. N., Schreiter, E. R., Zou, P., Wiley, H. S., Ting, A. Y., Lee, R. T., & Lauffenburger, D. A. (2010). The heparin-binding domain of HB-EGF mediates localization to sites of cell-cell contact and prevents HB-EGF proteolytic release. *Journal of Cell Science, 123*(Pt 13), 2308–2318.

Raman, R., Sasisekharan, V., & Sasisekharan, R. (2005). Structural insights into biological roles of protein-glycosaminoglycan interactions. *Chemistry & Biology, 12*(3), 267–277.

Ruhrberg, C., Gerhardt, H., Golding, M., Watson, R., Ioannidou, S., Fujisawa, H., et al. (2002). Spatially restricted patterning cues provided by heparin-binding VEGF-A control blood vessel branching morphogenesis. *Genes and Development, 16*(20), 2684–2698.

Rutkowski, J. M., Moya, M., Johannes, J., Goldman, J., & Swartz, M. A. (2006). Secondary lymphedema in the mouse tail: Lymphatic hyperplasia, VEGF-C upregulation, and the protective role of MMP-9. *Microvascular Research, 72*(3), 161–171.

Sabine, A., Agalarov, Y., Maby-El Hajjami, H., Jaquet, M., Hagerling, R., Pollmann, C., et al. (2012). Mechanotransduction, PROX1, and FOXC2 cooperate to control connexin37 and calcineurin during lymphatic-valve formation. *Developmental Cell, 22*(2), 430–445.

Saharinen, P., Helotera, H., Miettinen, J., Norrmen, C., D'Amico, G., Jeltsch, M., et al. (2010). Claudin-like protein 24 interacts with the VEGFR-2 and VEGFR-3 pathways and regulates lymphatic vessel development. *Genes and Development, 24*(9), 875–880.

Stratman, A. N., Malotte, K. M., Mahan, R. D., Davis, M. J., & Davis, G. E. (2009). Pericyte recruitment during vasculogenic tube assembly stimulates endothelial basement membrane matrix formation. *Blood, 114*(24), 5091–5101.

Tammela, T., He, Y., Lyytikka, J., Jeltsch, M., Markkanen, J., Pajusola, K., et al. (2007). Distinct architecture of lymphatic vessels induced by chimeric vascular endothelial growth factor-C/vascular endothelial growth factor heparin-binding domain fusion proteins. *Circulation Research, 100*(10), 1468–1475.

Vainionpaa, N., Butzow, R., Hukkanen, M., Jackson, D. G., Pihlajaniemi, T., Sakai, L. Y., & Virtanen, I. (2007). Basement membrane protein distribution in LYVE-1-immunoreactive lymphatic vessels of normal tissues and ovarian carcinomas. *Cell and Tissue Research, 328*(2), 317–328.

Wiig, H., Keskin, D., & Kalluri, R. (2010). Interaction between the extracellular matrix and lymphatics: Consequences for lymphangiogenesis and lymphatic function. *Matrix Biology, 29*(8), 645–656.

Yin, X., Johns, S. C., Lawrence, R., Xu, D., Reddi, K., Bishop, J. R., et al. (2011). Lymphatic endothelial heparan sulfate deficiency results in altered growth responses to vascular endothelial growth factor-C (VEGF-C). *Journal of Biological Chemistry, 286*(17), 14952–14962.

Chapter 6
Interplay of Mechanotransduction, FOXC2, Connexins, and Calcineurin Signaling in Lymphatic Valve Formation

Amélie Sabine and Tatiana V. Petrova

Abstract The directional flow of lymph is maintained by hundreds of intraluminal lymphatic valves. Lymphatic valves are crucial to prevent lymphedema, accumulation of fluid in the tissues, and to ensure immune surveillance; yet, the mechanisms of valve formation are only beginning to be elucidated. In this chapter, we will discuss the main steps of lymphatic valve morphogenesis, the important role of mechanotransduction in this process, and the genetic program regulated by the transcription factor Foxc2, which is indispensable for all steps of valve development. Failure to form mature collecting lymphatic vessels and valves causes the majority of postsurgical lymphedema, e.g., in breast cancer patients. Therefore, this knowledge will be useful for diagnostics and development of better treatments of secondary lymphedema.

6.1 Introduction

Lymphatic vessels contain two distinct functional compartments: lymphatic capillaries and collecting vessels. The main function of lymphatic capillaries is the uptake of interstitial fluid. They also serve as a principal point of entry for antigen-presenting cells, which is important for triggering immune responses.

A. Sabine
Department of Oncology, CHUV-UNIL, Ch. des Boveresses 155, CH-1066 Epalinges, Switzerland

T.V. Petrova (✉)
Department of Oncology, CHUV-UNIL, Ch. des Boveresses 155, CH-1066 Epalinges, Switzerland

Department of Biochemistry, University of Lausanne, Ch. des Boveresses 155, CH-1066 Epalinges, Switzerland

École Polytechnique Fédérale de Lausanne (EPFL), Swiss Institute for Experimental Cancer Research (ISREC), Lausanne, Switzerland
e-mail: tatiana.petrova@unil.ch

F. Kiefer and S. Schulte-Merker (eds.), *Developmental Aspects of the Lymphatic Vascular System*, Advances in Anatomy, Embryology and Cell Biology 214,
DOI 10.1007/978-3-7091-1646-3_6,

Collecting lymphatic vessels receive the lymph from capillaries and transport it towards lymph nodes. Efficient transport of lymph is enabled through several mechanisms: in capillaries overlapping endothelial cell borders form flaps, which prevent lymph exit immediately after the uptake; contraction of smooth muscle cells and surrounding tissues, such as skeletal muscle, propels the lymph in collecting vessels; and intraluminal valves prevent lymph reflux.

The exact number of lymphatic valves in humans or mice is not known. Adult mouse mesenteric lymphatic vessels have at least 500–800 valves (E. Bovay and A. Sabine, unpublished observation), highlighting the importance of this system for normal lymph transport. In this chapter, we will review the molecular mechanisms underlying formation of lymphatic valves with an emphasis on the role of forkhead transcription factor Foxc2, which acts as an essential regulator of valve development. In addition, current knowledge of human pathologies linked to dysfunctional or malformed lymphatic valves will be presented.

6.2 Lymphatic Valve Structure and Function

Lymphatic valves are intraluminal, bicuspid structures positioned at regular intervals in collecting lymphatic vessels (Fig. 6.1). The area between two valves, called lymphangion, is overlaid with smooth muscle cells, whereas the valve site is devoid of smooth muscle cells (Gnepp and Green 1980; Bouvree et al. 2012; Jurisic et al. 2012). Valve leaflets are covered by a layer of endothelial cells on each side, and the inner part of the leaflet is composed of a specialized extracellular matrix core.

Valve endothelial cells produce high levels of transcription factors Prox1, Foxc2, Nfatc1, and Gata2, cell adhesion molecules and receptor tyrosine kinases integrin α9, VEGFR-2, VEGFR-3, and ephrinB2, and extracellular matrix proteins laminin α5, collagen IV, and a splice isoform of fibronectin, fibronectin-EIIIA, and of podocalyxin, a major glycocalyx constituent (Petrova et al. 2004; Bazigou et al. 2009a; Norrmen et al. 2009; Kazenwadel et al. 2012; Sabine et al. 2012). Valve endothelial cells also strongly bind *Lycopersicon esculentum* lectin, which interacts with glycoconjugates in the mouse endothelial glycocalyx (Tammela et al. 2007). A number of valve-specific markers, such as laminin α5, collagen IV, and podocalyxin, are also highly produced by the blood vasculature. Lymphatic endothelial-specific podoplanin and VEGFR-3 continue to be produced by the valve endothelium, whereas levels of other typical lymphatic endothelial markers, such as Lyve-1 and neuropilin-2 (Nrp2), are low (Makinen et al. 2005; Bouvree et al. 2012; Sabine et al. 2012). Venous valves express high levels of Prox1, integrin α9, and ephrinB2 (Bazigou et al. 2011), and Prox1 is also expressed in cardiac valves (Wilting et al. 2002). This argues that valve endothelial cells represent a highly specialized subset of endothelial cells, characterized by a unique combination of both blood and lymphatic endothelial expression profiles.

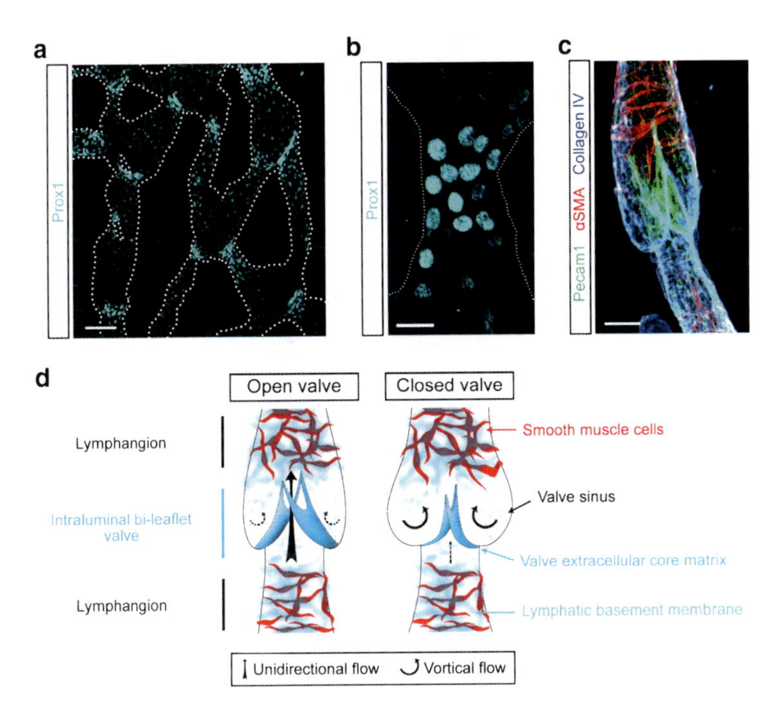

Fig. 6.1 Structure and function of lymphatic valves. (**a**) Regular positioning of valves in dorsal skin lymphatic vessels. Staining of E17.5 skin for Prox1 (*blue*). *Dashed lines* outline lymphatic vessels. *Scale bar*: 100 μm. (**b**) Early Prox1^{high} lymphatic valve-forming cells detected at E16.0 in the mesentery (*blue*). *Scale bar*: 20 μm. (**c**) Adult ear lymphatic valve: staining for the endothelial marker Pecam-1 (*green*), the vascular smooth muscle marker αSMA (*red*), and the basement membrane marker collagen IV (*purple*). *Scale bar*: 100 μm. (**d**) Schematic view of the structure of lymphatic valves and their function. When lymph accumulates in a lymphangion, smooth muscle cells constrict in response to the increase in luminal pressure and propel lymph through the opening valve into the following segment. Closure of the valve prevents lymph backflow. Lymphatic valve sites are devoid of smooth muscle cells to allow for leaflet movements and sinus deformation during the valve opening/closure cycle

The distinct molecular makeup of valve cells is likely a result of their exposure to complex fluid patterns (reviewed in (Chiu and Chien 2011; Bazigou and Makinen 2013)). Such conditions in combination with specialized extracellular matrix composition are likely important for a valve-specific transcriptional program. Interestingly, aortic and aortic valve endothelial cells conserve some of their unique in vivo features in culture, such as specific gene expression profiles and differential alignment in response to fluid flow (Holliday et al. 2011; Butcher et al. 2006; Butcher et al. 2004), but this question was not studied in case of lymphatic valve cells.

6.3 Steps of Lymphatic Valve Morphogenesis

Based on the analysis of mouse mesenteric lymphatics, the formation of lymphatic valves can be divided in the following stages: (1) valve initiation, (2) formation of a circumferential valve territory, and (3) valve leaflet elongation in the lumen and maturation (Sabine et al. 2012) (Fig. 6.2a). The earliest sign of valve formation is a localized upregulation of transcription factors Prox1 and Foxc2, accompanied by decreased expression of Nrp2 and Lyve-1. Early lymphatic valve-forming cells (LVCs) adopt a cuboidal shape, with characteristic round nuclei, distinct from the surrounding more elongated lymphangion endothelial cells, and thus define the valve territory (Fig. 6.1b). In the subsequent steps, LVCs elongate, reorientate, and invaginate into the vessel lumen, forming bicuspid leaflets (reviewed in (Bazigou and Makinen 2013; Koltowska et al. 2013; Tatin et al. 2013)).

In the past few years, a number of molecular regulators of lymphatic valve development have been identified, and the corresponding mouse models are listed in Table 6.1. The role of extracellular matrix components and cell adhesion molecules in valve development is reviewed in Chap. 5 in this book. Here we will concentrate on Foxc2 transcription factor and its co- and downstream effectors, important for lymphatic valve morphogenesis.

6.3.1 FOXC2: An Essential Regulator of Lymphatic Valve Development

Foxc2 belongs to the family of forkhead transcription factors and it contains a conserved 100-amino-acid DNA-binding domain called forkhead or winged helix, flanked on both sides by transcriptional activation domains (reviewed in (Benayoun et al. 2011; Kume 2008)). Foxc2 phosphorylation on eight conserved Ser/Thr residues is important for enhancing the interaction of Foxc2 with the DNA in the context of native chromatin (Ivanov et al. 2013). Foxc2 also contains two sumoylation sites and this modification leads to the reduction of Foxc2 transcriptional activity (Danciu et al. 2012). Thus, the interplay of sumoylation and phosphorylation potentially fine-tunes Foxc2 activity.

During development, Foxc2 is expressed in many tissues, derived from mesoderm and neural crest (reviewed in (Kume 2008)). In the cardiovascular system Foxc2 is highly expressed in arterial smooth muscle and endothelial cells (Kume et al. 2001). In the lymphatic vasculature Foxc2 levels are low in capillaries, whereas Foxc2 is highly expressed in lymphatic valves, with intermediate expression in lymphangions (Petrova et al. 2004; Norrmen et al. 2009).

Mice with germline inactivation of Foxc2 die perinatally because of inability to breathe (Kume et al. 2001; Iida et al. 1997). In addition, they have incomplete development of the aorta and defective skeletal and kidney development (Iida et al. 1997; Takemoto et al. 2006; Kume et al. 2000). Defects of the lymphatic

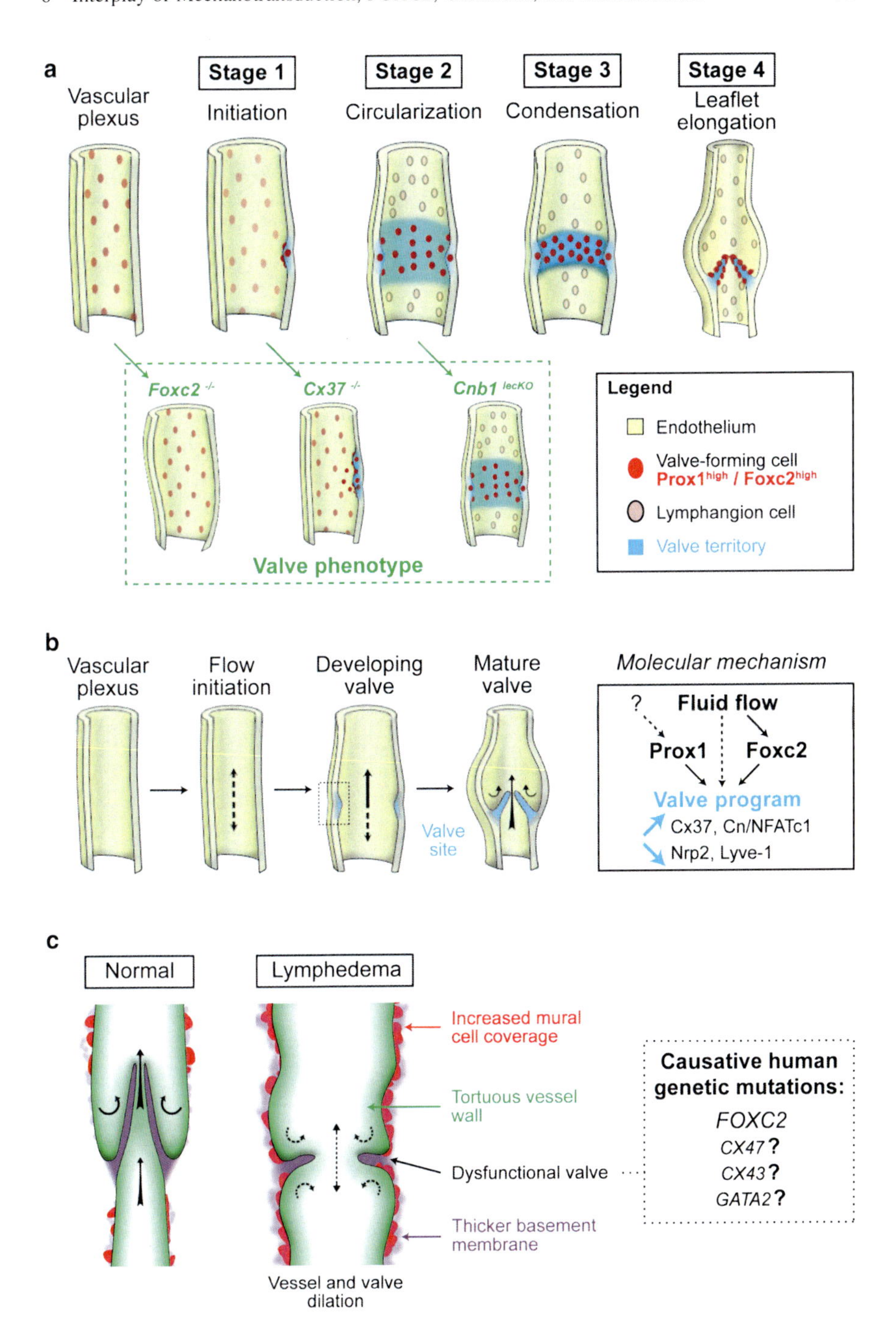

Fig. 6.2 Mechanisms of lymphatic valve development and dysfunction. (**a**) Current view of lymphatic valve developmental stages. Inactivation of Foxc2 prevents valve initiation, whereas loss of Cx37 or CnB1 impairs formation of ring-like valve territory or reorientation of LVCs (lecKO, deletion specific for lymphatic endothelial cells). Additional valve phenotypes are discussed in

vasculature include agenesis of lymphatic valves and failure to form mature collecting vessels. In the absence of Foxc2, presumptive collecting vessels are not able to remodel and to downregulate the expression of lymphatic capillary markers Lyve-1, VEGFR-3, and Ccl21. Interestingly, in *Foxc2*$^{-/-}$ embryos, lymphatic capillaries, normally devoid of smooth muscle cells and basement membrane, acquire ectopic mural cell and basement membrane coverage. This feature may account for the decreased lymphatic sprouting and lymphatic hyperplasia observed in *Foxc2*$^{-/-}$ mice (Petrova et al. 2004; Norrmen et al. 2009; Kriederman et al. 2003).

In humans, heterozygous loss-of-function mutations in *FOXC2* cause lymphedema-distichiasis (LD), a late-onset lymphedema, which is also characterized by abnormal development of sebum-producing Meibomian glands in the eye (Fang et al. 2000). Lymphatic vessels are present or even hyperplastic in LD patients, but lymph transport is inefficient. LD patients also frequently develop venous reflux and varicose veins, indicating a compromised function of venous valves (Mellor et al. 2007).

6.3.2 *Ca*$^{2+}$*/Calcineurin/NFAT Signaling in Valve Morphogenesis*

Large-scale analysis of Foxc2 chromosome occupancy in lymphatic endothelial cells identified over two thousand binding sites, which are mostly located in distal enhancer regions (Norrmen et al. 2009). Bioinformatics analysis of Foxc2 ChIP sequences revealed the enrichment for nuclear factor of activated T-cell (NFAT) binding sites as well, suggesting that Ca^{2+}/calcineurin/NFAT signaling pathway cooperates with Foxc2 signaling in lymphatic vessels (Norrmen et al. 2009; Sabine et al. 2012). Calcineurin is a ubiquitously expressed Ca^{2+}-calmodulin-activated serine/threonine phosphatase. It is a heterodimer composed of a calmodulin-binding catalytic subunit A (CnA) and a Ca^{2+}-binding regulatory subunit B1 (CnB1) in most tissues. Increase in intracellular Ca^{2+} leads to activation of calcineurin phosphatase activity, dephosphorylation of NFAT transcription factors, and their translocation from the cytosol to the nucleus. The NFAT family contains five members, of which NFAT1-NFAT4 are Ca^{2+}/calcineurin-dependent, whereas NFAT5 is regulated by osmotic stress (Hogan et al. 2003). At least NFAT2 (also known as NFATc1) is highly expressed in lymphatic endothelial cells in vitro (Norrmen et al. 2009; Kulkarni et al. 2009), and it is localized in the nuclei of

Fig. 6.2 (continued) Chap. 5. (**b**) Role of flow in lymphatic valve development. *Arrows* indicate the direction of flow. (**c**) Normal and dysfunctional lymphatic collecting vessel. *Arrows* indicate the direction of flow. Collecting vessel dilation, abnormal coverage of valve site with smooth muscle cells and basement membrane, or shortening of valve leaflets impairs normal lymphatic function and will lead to lymphedema

Table 6.1 Mouse mutants displaying defects of lymphatic valve formation

Gene symbol	Function	Lymphatic valve phenotype	Additional lymphatic phenotypes
Growth factors, receptors, and intracellular signaling components			
Akt1	Ser/Thr kinase, PI3 kinase signaling pathway	Valve agenesis in small collecting lymphatic vessels (Zhou et al. 2010)	Capillary hypoplasia, dilation and decreased SMC coverage of small collecting lymphatic vessels
Angpt2	Growth factor, ligand of Tie receptor tyrosine kinases	Valve agenesis (Gale et al. 2002; Dellinger et al. 2008)	Hypoplasia, defective remodeling, chylous ascites, ectopic smooth muscle cell coverage
Pi3kr1	Regulatory subunits of class IA PI3 kinases	Valve agenesis (Mouta-Bellum et al. 2009)	Chylous ascites, intestinal lymphangiectasia, impaired sprouting
Cnb1	Calcineurin subunit, Ca^{2+}/calcineurin/NFAT signaling	Arrested development of lymphatic valves (Sabine et al. 2012)	–
Efnb2 (mutation of PDZ binding site)	Ligand of EphB receptor tyrosine kinases	Agenesis of lymphatic valves, retrograde lymph flow (Makinen et al. 2005)	Impaired sprouting of capillaries, ectopic mural cells, chylothorax
ECM components and receptors			
Itgα9	Adhesion, assembly of ECM fibrils	Arrested development of lymphatic valves, retrograde lymph flow (Bazigou et al. 2009a)	Chylothorax
Fn1 (removal of EDA domain)	ECM component, ligand of Itgα9	Arrested development of lymphatic valves, reduced number of lymphatic valves at birth, recovery in adult mice (−/−) (Bazigou et al. 2009a)	–
Transcription factors			
Foxc2	Transcription factor	Agenesis of lymphatic valves, retrograde lymph flow (Petrova et al. 2004; Dagenais et al. 2004; Norrmen et al. 2009)	Impaired patterning of capillaries, ectopic mural cells
Guidance molecules			
Sema3a	Ligand of Nrp1 and Plexin A1, repulsive guidance cues	Small lymphatic valves, abnormal lymph drainage (Bouvree et al. 2012)	Ectopic coverage of valve area with smooth muscle cells, also in mice treated with antibodies blocking interaction of Nrp1 with Sema3a (Bouvree

(continued)

Table 6.1 (continued)

Gene symbol	Function	Lymphatic valve phenotype	Additional lymphatic phenotypes
			et al. 2012; Jurisic et al. 2012)
Nrp1	Coreceptor of Sema3a, Sema3c, and VEGF	Small lymphatic valves (*Nrp1*$^{sema-/-}$) (Bouvree et al. 2012)	Ectopic coverage of valve area with smooth muscle cells
Plxna1	Receptor of Sema3a	Small lymphatic valves (Bouvree et al. 2012)	–
Gap junctions			
Connexin 37 (*Gja4*)	Gap junction protein	Arrested development of lymphatic valves, retrograde lymph flow (Kanady et al. 2011; Sabine et al. 2012)	Lymphatic hyperplasia
Connexin 43 (*Gja1*)	Gap junction protein	Agenesis of lymphatic valves (Kanady et al. 2011)	–

both developing and mature valves, indicating active calcineurin signaling (Norrmen et al. 2009; Sabine et al. 2012). Treatment with a pharmacological inhibitor of calcineurin, cyclosporine A (CsA), or lymphatic endothelial-specific inactivation of CnB1 prevents formation of lymphatic valve leaflets (Kulkarni et al. 2009; Norrmen et al. 2009). CsA-treated and *Nfatc1*$^{-/-}$ embryos also display abnormal lymphatic capillary patterning; however, the cell-autonomous vs. non-autonomous contribution of Ca^{2+}/calcineurin signaling to these defects remains to be investigated (Sabine et al. 2012; Norrmen et al. 2009).

Calcineurin signaling regulates early stages of mouse and zebrafish heart valve development and endocardial valve cells express high levels of Prox1 and nuclear NFATc1 (Wilting et al. 2002; de la Pompa et al. 1998; Chang et al. 2004; Beis et al. 2005). Thus, some common molecular pathways are shared by both heart and lymphatic valves. However, there may also be important differences. For example, early steps of mammalian heart valve formation involve endothelial-to-mesenchymal transition (EMT) in a subset of endocardial cushion cells and extensive valve cell proliferation (Hinton and Yutzey 2011), whereas EMT and LVC proliferation do not appear to be a prominent feature of lymphatic valve development ((Bazigou et al. 2009b) and our unpublished observation).

6.3.3 *Connexins and Lymphatic Valve Development*

Connexins are a large family of transmembrane proteins, which form gap junctions, a specific type of channel that allows direct transfer of small molecular weight compounds, such as ions, second messengers, or metabolites, between neighboring cells. Signaling via gap junctions synchronizes several multicellular responses, including collective cell migration, cardiomyocyte contraction, or hormone secretion from the pancreas (reviewed in (Wei et al. 2004)). In addition, hemichannels can also release small cytoplasmic molecules, such as ATP, into the extracellular compartment. Signaling via hemichannels contributes to many physiological and pathophysiological events, such as atherosclerosis (Pfenniger et al. 2013).

Lymphatic endothelial cells express connexin 37, connexin 43, and connexin 47 (Kanady et al. 2011; Sabine et al. 2012). During lymphatic valve formation, Cx37 expression is upregulated in the early LVCs (Sabine et al. 2012). Later, both Cx37 and Cx43 are found in developing and mature lymphatic valves in a non-overlapping fashion: Cx37 is predominantly located in the valve sinuses, whereas Cx43 is localized on the opposite side of the leaflet (Kanady et al. 2011; Sabine et al. 2012). Staining for Cx47 was detected in a small subset of Cx43-expressing valve cells (Kanady et al. 2011). *Cx37*$^{-/-}$ mice have a decreased number of lymphatic valves both during development and in adulthood (Kanady et al. 2011; Sabine et al. 2012). *Cx43*$^{-/-}$ embryos failed to develop any lymphatic valves (Kanady et al. 2011). Various combinations of Cx37 and Cx43 allele loss lead to exacerbation of the lymphatic vascular phenotype, e.g., chylothorax and premature death in adult *Cx37*$^{-/-}$ and *Cx43*$^{+/-}$mice (Kanady et al. 2011). Mechanistically, in the absence of Cx37, LVCs are induced, but do not form a well-organized valve territory (Sabine et al. 2012). Expression of Cx37 is jointly regulated by shear stress and Foxc2 in lymphatic endothelial cells in vitro (see below) and Cx37 is significantly decreased in *Foxc2*$^{-/-}$ mice, suggesting that Cx37 is one of Foxc2 target genes (Kanady et al. 2011; Sabine et al. 2012).

Heterozygous missense mutations in connexin 47 (GJC2) were identified as a cause of hereditary lymphedema (Ostergaard et al. 2011; Ferrell et al. 2010). A recent study also demonstrated the existence of clinically silent GJC2 mutations, which lead to lymphedema only after lymphatic vessels are damaged, following axillary lymph node dissection in breast cancer patients (Finegold et al. 2012). These data underscore the important role of genetic factors in the so far poorly understood predisposition for the development of acquired lymphedema, which affects up to 30 % of such patients.

Mutations in connexin 43 (*GJA1*) cause oculodentodigital syndrome (ODD; OMIM 164200), which most commonly affects the face, eyes, dentition, and digits. Recently, an ODD patient with a late-onset lymphedema was described, suggesting that lymphatic vascular dysfunction is an associated feature of ODD (Brice et al. 2013). The mutation in the patient with lymphedema was located in the second extracellular loop of connexin 43, previously shown to be important for the formation of functional gap junctions. Lymphoscintigraphy analysis further demonstrated impaired lymphatic drainage despite the normal visualization of lymphatic vessels, which is consistent with lymphatic valve defects rather than capillary hypoplasia (Brice et al. 2013). Mouse and human genetic studies demonstrate that specific connexins play important roles in lymphatic vascular patterning during development. However, the in vivo role of gap junctional communication vs. other functions of connexins still remains to be investigated.

6.4 Cooperation of Mechanotransduction and Transcriptional Regulation in the Valve Development Program

Lymphatic valves frequently develop at sites of lymphatic vessel branching and bifurcations (Kampmeier 1928; Sabine et al. 2012) (Fig. 6.1a), suggesting that the local disturbed flow plays an important role in valve development. Indeed, the valve initiation period coincides with the establishment of active lymphatic drainage in mouse embryos (Sabine et al. 2012). Early LVCs also display a cuboidal morphology, which parallels the observations in blood vessels, where endothelial cells located in the straight portion are elongated and aligned in the longitudinal axis of the vessel, whereas they have a polygonal morphology in branching regions, exposed to disturbed flow (reviewed in (Chiu and Chien 2011)). During zebrafish heart development, reversing flow initiates formation of cardiac valve leaflets through flow-induced transcription factor klf2a (Vermot et al. 2009). Cultured LECs subjected to oscillatory shear stress display many in vivo characteristics of early LVCs, such as high expression of Foxc2 and Cx37, activation of calcineurin/NFAT signaling, as well as acquisition of cuboidal cell shape and rearrangement of the actin cytoskeleton (Sabine et al. 2012). In contrast, no in vitro models are available to study subsequent steps of leaflet formation, which include LVC elongation, reorientation, and collective migration in the lumen, recently described in vivo (Tatin et al. 2013). Expression of Prox1 in lymphatic endothelial cells is repressed by high laminar shear stress (Chen et al. 2012), further implicating fluid flow in the control of lymphatic vessel function and differentiation.

In vitro knockdown of Cx37 prevented nuclear translocation of NFATc1 in response to shear stress, suggesting that gap junctional communication maintains the field of valve cells with high Ca^{2+}/calcineurin activity (Sabine et al. 2012). IP3, which regulates Ca^{2+} release from the intracellular stores, is a potential mediator of such cell synchronization; however, alternative mechanisms, such as ATP release via Cx37 hemichannels, cannot be excluded. Importantly, shear stress-induced expression of Cx37, nuclear translocation of NFATc1 in LECs, and changes in actin cytoskeleton require FOXC2 and PROX1 (Sabine et al. 2012). Thus, an emerging unifying concept is that input of both mechanical forces and valve-specific transcription factors is required for a full activation of the valve development program (Fig. 6.2b). Such interplay of extrinsic factors and intrinsic genetic programs is probably a prerequisite to determine the position and size of an individual lymphatic valve and to allow optimal adaptation of the vascular network growth to the increasing need of lymph transport. Conversely, disturbances in the formation or function of lymphatic valves will impair lymphatic drainage and will lead to the development of lymphedema (Fig. 6.2c).

6.5 Concluding Remarks

Lymphatic valves are one of the defining features of collecting lymphatic vessels and they are a prerequisite for efficient flow of the lymph. Human molecular genetics and mouse models have already provided a substantial amount of data on the molecular regulation of lymphatic valve morphogenesis; however, further studies are necessary to improve our understanding of this complex process. How do non-endothelial cells contribute to valve morphogenesis? What are further similarities and differences between cardiac, venous, and lymphatic valve development? Could known molecular targets and pathways be exploited to enhance the regeneration of valves in human lymphatic vessels? These and many other questions still await further investigation.

Acknowledgments We apologize to colleagues whose work could not be cited, because of space limitations. The research from T. Petrova laboratory is supported by the Swiss National Science Foundation (PPP0033-114898 and CRSII3_141811), Leenaards Foundation, Gebert Rüf Foundation, and EU FP7 ITN-2012-317250.

References

Bazigou, E., Lyons, O. T., Smith, A., Venn, G. E., Cope, C., Brown, N. A., & Makinen, T. (2011). Genes regulating lymphangiogenesis control venous valve formation and maintenance in mice. *Journal of Clinical Investigation, 121*(8), 2984–2992.

Bazigou, E., & Makinen, T. (2013). Flow control in our vessels: Vascular valves make sure there is no way back. *Cellular and Molecular Life Sciences, 70*(6), 1055–1066.

Bazigou, E., Xie, S., Chen, C., Weston, A., Miura, N., Sorokin, L., et al. (2009a). Integrin-alpha9 is required for fibronectin matrix assembly during lymphatic valve morphogenesis. *Developmental Cell, 17*(2), 175–186.

Bazigou, E., Xie, S., Chen, C., Weston, A., Miura, N., Sorokin, L., et al. (2009b). Integrin-alpha9 is required for fibronectin matrix assembly during lymphatic valve morphogenesis. *Developmental Cell, 17*(2), 175–186. doi:10.1016/j.devcel.2009.06.017.

Beis, D., Bartman, T., Jin, S. W., Scott, I. C., D'Amico, L. A., Ober, E. A., et al. (2005). Genetic and cellular analyses of zebrafish atrioventricular cushion and valve development. *Development, 132*(18), 4193–4204.

Benayoun, B. A., Caburet, S., & Veitia, R. A. (2011). Forkhead transcription factors: Key players in health and disease. *Trends in Genetics, 27*(6), 224–232.

Bouvree, K., Brunet, I., Del Toro, R., Gordon, E., Prahst, C., Cristofaro, B., et al. (2012). Semaphorin3A, Neuropilin-1, and PlexinA1 are required for lymphatic valve formation. *Circulation Research, 111*(4), 437–445.

Brice, G., Ostergaard, P., Jeffery, S., Gordon, K., Mortimer, P., & Mansour, S. (2013). A novel mutation in GJA1 causing oculodentodigital syndrome and primary lymphoedema in a three generation family. *Clinical Genetics, 64*(4), 378–381.

Butcher, J. T., Penrod, A. M., Garcia, A. J., & Nerem, R. M. (2004). Unique morphology and focal adhesion development of valvular endothelial cells in static and fluid flow environments. *Arteriosclerosis, Thrombosis, and Vascular Biology, 24*(8), 1429–1434.

Butcher, J. T., Tressel, S., Johnson, T., Turner, D., Sorescu, G., Jo, H., & Nerem, R. M. (2006). Transcriptional profiles of valvular and vascular endothelial cells reveal phenotypic differences: Influence of shear stress. *Arteriosclerosis, Thrombosis, and Vascular Biology, 26*(1), 69–77.

Chang, C. P., Neilson, J. R., Bayle, J. H., Gestwicki, J. E., Kuo, A., Stankunas, K., et al. (2004). A field of myocardial-endocardial NFAT signaling underlies heart valve morphogenesis. *Cell, 118*(5), 649–663.

Chen, C. Y., Bertozzi, C., Zou, Z., Yuan, L., Lee, J. S., Lu, M., et al. (2012). Blood flow reprograms lymphatic vessels to blood vessels. *Journal of Clinical Investigation, 122*(6), 2006–2017.

Chiu, J. J., & Chien, S. (2011). Effects of disturbed flow on vascular endothelium: Pathophysiological basis and clinical perspectives. *Physiological Reviews, 91*(1), 327–387.

Dagenais, S. L., Hartsough, R. L., Erickson, R. P., Witte, M. H., Butler, M. G., & Glover, T. W. (2004). Foxc2 is expressed in developing lymphatic vessels and other tissues associated with lymphedema-distichiasis syndrome. *Gene Expression Patterns, 4*(6), 611–619.

Danciu, T. E., Chupreta, S., Cruz, O., Fox, J. E., Whitman, M., & Iniguez-Lluhi, J. A. (2012). Small ubiquitin-like modifier (SUMO) modification mediates function of the inhibitory domains of developmental regulators FOXC1 and FOXC2. *Journal of Biological Chemistry, 287*(22), 18318–18329.

de la Pompa, J. L., Timmerman, L. A., Takimoto, H., Yoshida, H., Elia, A. J., Samper, E., et al. (1998). Role of the NF-ATc transcription factor in morphogenesis of cardiac valves and septum. *Nature, 392*(6672), 182–186.

Dellinger, M., Hunter, R., Bernas, M., Gale, N., Yancopoulos, G., Erickson, R., & Witte, M. (2008). Defective remodeling and maturation of the lymphatic vasculature in Angiopoietin-2 deficient mice. *Developmental Biology, 319*(2), 309–320.

Fang, J., Dagenais, S. L., Erickson, R. P., Arlt, M. F., Glynn, M. W., Gorski, J. L., et al. (2000). Mutations in FOXC2 (MFH-1), a forkhead family transcription factor, are responsible for the hereditary lymphedema-distichiasis syndrome. *American Journal of Human Genetics, 67*(6), 1382–1388.

Ferrell, R. E., Baty, C. J., Kimak, M. A., Karlsson, J. M., Lawrence, E. C., Franke-Snyder, M., et al. (2010). GJC2 missense mutations cause human lymphedema. *American Journal of Human Genetics, 86*(6), 943–948.

Finegold, D. N., Baty, C. J., Knickelbein, K. Z., Perschke, S., Noon, S. E., Campbell, D., et al. (2012). Connexin 47 mutations increase risk for secondary lymphedema following breast cancer treatment. *Clinical Cancer Research, 18*(8), 2382–2390.

Gale, N. W., Thurston, G., Hackett, S. F., Renard, R., Wang, Q., McClain, J., et al. (2002). Angiopoietin-2 is required for postnatal angiogenesis and lymphatic patterning, and only the latter role is rescued by Angiopoietin-1. *Developmental Cell, 3*(3), 411–423.

Gnepp, D. R., & Green, F. H. (1980). Scanning electron microscopic study of canine lymphatic vessels and their valves. *Lymphology, 13*(2), 91–99.

Hinton, R. B., & Yutzey, K. E. (2011). Heart valve structure and function in development and disease. *Annual Review of Physiology, 73*, 29–46.

Hogan, P. G., Chen, L., Nardone, J., & Rao, A. (2003). Transcriptional regulation by calcium, calcineurin, and NFAT. *Genes and Development, 17*(18), 2205–2232.

Holliday, C. J., Ankeny, R. F., Jo, H., & Nerem, R. M. (2011). Discovery of shear- and side-specific mRNAs and miRNAs in human aortic valvular endothelial cells. *American Journal of Physiology - Heart and Circulatory Physiology, 301*(3), H856–H867.

Iida, K., Koseki, H., Kakinuma, H., Kato, N., Mizutani-Koseki, Y., Ohuchi, H., et al. (1997). Essential roles of the winged helix transcription factor MFH-1 in aortic arch patterning and skeletogenesis. *Development, 124*(22), 4627–4638.

Ivanov, K. I., Agalarov, Y., Valmu, L., Samuilova, O., Liebl, J., Houhou, N., et al. (2013). Phosphorylation regulates FOXC2-mediated transcription in lymphatic endothelial cells. *Molecular and Cellular Biology, 33*(19), 3749–3761.

Jurisic, G., Maby-El Hajjami, H., Karaman, S., Ochsenbein, A. M., Alitalo, A., Siddiqui, S. S., et al. (2012). An unexpected role of semaphorin3a-neuropilin-1 signaling in lymphatic vessel maturation and valve formation. *Circulation Research, 111*(4), 426–436.

Kampmeier, O. F. (1928). The genetic history of the valves in the lymphatic system of man. *The American Journal of Anatomy, 40*(3), 413–457.

Kanady, J. D., Dellinger, M. T., Munger, S. J., Witte, M. H., & Simon, A. M. (2011). Connexin37 and Connexin43 deficiencies in mice disrupt lymphatic valve development and result in lymphatic disorders including lymphedema and chylothorax. *Developmental Biology, 354* (2), 253–266.

Kazenwadel, J., Secker, G. A., Liu, Y. J., Rosenfeld, J. A., Wildin, R. S., Cuellar-Rodriguez, J., et al. (2012). Loss-of-function germline GATA2 mutations in patients with MDS/AML or MonoMAC syndrome and primary lymphedema reveal a key role for GATA2 in the lymphatic vasculature. *Blood, 119*(5), 1283–1291.

Koltowska, K., Betterman, K. L., Harvey, N. L., & Hogan, B. M. (2013). Getting out and about: The emergence and morphogenesis of the vertebrate lymphatic vasculature. *Development, 140* (9), 1857–1870.

Kriederman, B. M., Myloyde, T. L., Witte, M. H., Dagenais, S. L., Witte, C. L., Rennels, M., et al. (2003). FOXC2 haploinsufficient mice are a model for human autosomal dominant lymphedema-distichiasis syndrome. *Human Molecular Genetics, 12*(10), 1179–1185.

Kulkarni, R. M., Greenberg, J. M., & Akeson, A. L. (2009). NFATc1 regulates lymphatic endothelial development. *Mechanisms of Development, 126*(5–6), 350–365.

Kume, T. (2008). Foxc2 transcription factor: A newly described regulator of angiogenesis. *Trends in Cardiovascular Medicine, 18*(6), 224–228.

Kume, T., Deng, K., & Hogan, B. L. (2000). Murine forkhead/winged helix genes Foxc1 (Mf1) and Foxc2 (Mfh1) are required for the early organogenesis of the kidney and urinary tract. *Development, 127*(7), 1387–1395.

Kume, T., Jiang, H., Topczewska, J. M., & Hogan, B. L. (2001). The murine winged helix transcription factors, Foxc1 and Foxc2, are both required for cardiovascular development and somitogenesis. *Genes and Development, 15*(18), 2470–2482.

Makinen, T., Adams, R. H., Bailey, J., Lu, Q., Ziemiecki, A., Alitalo, K., et al. (2005). PDZ interaction site in ephrinB2 is required for the remodeling of lymphatic vasculature. *Genes and Development, 19*(3), 397–410.

Mellor, R. H., Brice, G., Stanton, A. W., French, J., Smith, A., Jeffery, S., et al. (2007). Mutations in FOXC2 are strongly associated with primary valve failure in veins of the lower limb. *Circulation, 115*(14), 1912–1920.

Mouta-Bellum, C., Kirov, A., Miceli-Libby, L., Mancini, M. L., Petrova, T. V., Liaw, L., et al. (2009). Organ-specific lymphangiectasia, arrested lymphatic sprouting, and maturation defects resulting from gene-targeting of the PI3K regulatory isoforms p85alpha, p55alpha, and p50alpha. *Developmental Dynamics, 238*(10), 2670–2679.

Norrmen, C., Ivanov, K. I., Cheng, J., Zangger, N., Delorenzi, M., Jaquet, M., et al. (2009). FOXC2 controls formation and maturation of lymphatic collecting vessels through cooperation with NFATc1. *Journal of Cell Biology, 185*(3), 439–457.

Ostergaard, P., Simpson, M. A., Brice, G., Mansour, S., Connell, F. C., Onoufriadis, A., et al. (2011). Rapid identification of mutations in GJC2 in primary lymphoedema using whole exome sequencing combined with linkage analysis with delineation of the phenotype. *Journal of Medical Genetics, 48*(4), 251–255.

Petrova, T. V., Karpanen, T., Norrmen, C., Mellor, R., Tamakoshi, T., Finegold, D., et al. (2004). Defective valves and abnormal mural cell recruitment underlie lymphatic vascular failure in lymphedema distichiasis. *Nature Medicine, 10*(9), 974–981.

Pfenniger, A., Chanson, M., & Kwak, B. R. (2013). Connexins in atherosclerosis. *Biochimica et Biophysica Acta, 1828*(1), 157–166.

Sabine, A., Agalarov, Y., Maby-El Hajjami, H., Jaquet, M., Hagerling, R., Pollmann, C., et al. (2012). Mechanotransduction, PROX1, and FOXC2 cooperate to control connexin37 and calcineurin during lymphatic-valve formation. *Developmental Cell, 22*(2), 430–445.

Takemoto, M., He, L., Norlin, J., Patrakka, J., Xiao, Z., Petrova, T., et al. (2006). Large-scale identification of genes implicated in kidney glomerulus development and function. *EMBO Journal, 25*(5), 1160–1174.

Tammela, T., Saaristo, A., Holopainen, T., Lyytikka, J., Kotronen, A., Pitkonen, M., et al. (2007). Therapeutic differentiation and maturation of lymphatic vessels after lymph node dissection and transplantation. *Nature Medicine, 13*(12), 1458–1466.

Tatin, F., Taddei, A., Weston, A., Fuchs, E., Devenport, D., Tissir, F., & Makinen, T. (2013). Planar cell polarity protein Celsr1 regulates endothelial adherens junctions and directed cell rearrangements during valve morphogenesis. *Developmental Cell, 26*(1), 31–44.

Vermot, J., Forouhar, A. S., Liebling, M., Wu, D., Plummer, D., Gharib, M., & Fraser, S. E. (2009). Reversing blood flows act through klf2a to ensure normal valvulogenesis in the developing heart. *PLoS Biology, 7*(11), e1000246.

Wei, C. J., Xu, X., & Lo, C. W. (2004). Connexins and cell signaling in development and disease. *Annual Review of Cell and Developmental Biology, 20*, 811–838.

Wilting, J., Papoutsi, M., Christ, B., Nicolaides, K. H., von Kaisenberg, C. S., Borges, J., et al. (2002). The transcription factor Prox1 is a marker for lymphatic endothelial cells in normal and diseased human tissues. *FASEB Journal, 16*(10), 1271–1273.

Zhou, F., Chang, Z., Zhang, L., Hong, Y. K., Shen, B., Wang, B., et al. (2010). Akt/Protein kinase B is required for lymphatic network formation, remodeling, and valve development. *American Journal of Pathology, 177*(4), 2124–2133.

Chapter 7
Development of Secondary Lymphoid Organs in Relation to Lymphatic Vasculature

Serge A. van de Pavert and Reina E. Mebius

Abstract Although the initial event in lymphatic endothelial specification occurs slightly before the initiation of lymph node formation in mice, the development of lymphatic vessels and lymph nodes occurs within the same embryonic time frame. Specification of lymphatic endothelial cells starts around embryonic day 10 (E10), followed by endothelial cell budding and formation of the first lymphatic structures. Through lymphatic endothelial cell sprouting these lymph sacs give rise to the lymphatic vasculature which is complete by E15.5 in mice. It is within this time frame that lymph node formation is initiated and the first structure is secured in place. As lymphatic vessels are crucially involved in the functionality of the lymph nodes, the recent insight that both structures depend on common developmental signals for their initiation provides a molecular mechanism for their coordinated formation. Here, we will describe the common developmental signals needed to properly start the formation of lymphatic vessels and lymph nodes and their interdependence in adult life.

7.1 Introduction

Lymphatic vessels are instrumental for the transport of exudates, which develop at capillary beds, back to the bloodstream. This fluid collects as lymph in lymphatic vessels and moves to the draining lymph nodes. Cells and small molecules derived from these peripheral sites are carried with the lymph to the draining lymph nodes.

S.A. van de Pavert
Royal Netherlands Academy of Arts and Sciences, Hubrecht Institute, Uppsalalaan 8, 3584CT Utrecht, The Netherlands
e-mail: s.vandepavert@hubrecht.eu

R.E. Mebius (✉)
Department of Molecular Cell Biology and Immunology, VU University Medical Center, van der Boechorststraat 7, 1081BT Amsterdam, The Netherlands
e-mail: r.mebius@vumc.nl

F. Kiefer and S. Schulte-Merker (eds.), *Developmental Aspects of the Lymphatic Vascular System*, Advances in Anatomy, Embryology and Cell Biology 214,
DOI 10.1007/978-3-7091-1646-3_7, © Springer-Verlag Wien 2014

In the absence of an infection this does not lead to measurable changes within these lymph nodes. However, upon infection, antigen-presenting cells will carry antigen derived from the infected area and travel with the lymph to the draining lymph nodes. Here they will present the antigen to lymphocytes, which will start the adaptive immune response. Hence, properly placed lymphatic vessels and lymph nodes are crucial for an adequate and fast immune response needed for protection against infectious agents that enter our body. It is therefore not surprising that both lymphatic vessels and lymph nodes are formed in a coordinated fashion, as lymphatic vessels and lymph nodes form within the same time frame, while sharing essential differentiation signals.

Here, we will discuss this coordinated process leading to the formation of a functional lymphatic system.

7.2 Lymph Node Formation

Lymph node (LN) formation requires a coordinated interaction between mesenchymal lymphoid tissue organizer cells (LTo) and hematopoietic lymphoid tissue inducer cells (LTi) (van de Pavert and Mebius 2010). In the course of this process, lymph node anlagen transform from clusters of LTo and LTi cells during embryogenesis to fully organized lymph nodes containing specific compartments for T- or B-cell activation after birth in mice (Cupedo and Mebius 2005). The initiating and crucial event during the formation of peripheral lymph nodes is the induction of the chemokine Cxcl13, produced by stromal cells. Expression of Cxcl13 in defined places results in the attraction of the first hematopoietic cells, the LTi cells, to the locations where lymph nodes will form (Ansel et al. 2000; Luther et al. 2003; van de Pavert et al. 2009). These LTi cells will subsequently start communicating with the stromal cells, which will secure the first cluster of hematopoietic cells in place and will lead to a further enlargement of the lymph node. To migrate here, LTi cells selectively express Cxcr5, the receptor for Cxcl13 (Cherrier et al. 2012; Honda et al. 2001; Luther et al. 2003; Mebius et al. 1997; van de Pavert et al. 2009). And thus the induction of Cxcl13 is the determining step for starting the formation of lymph nodes. In addition, LTi cells express Ccr7, the receptor for Ccl21, produced by lymphatic endothelial cells, as well as for Ccl19 (Ansel et al. 2000; Honda et al. 2001; van de Pavert et al. 2009). And indeed, expression of Ccr7 on LTi cells can also contribute to the first clustering of these cells, in those lymph nodes where lymphatic endothelial cells are already present.

The initial induction of Cxcl13 expression crucially depends on the vitamin A metabolic breakdown product, retinoic acid (van de Pavert et al. 2009). Retinoic acid binds to nuclear retinoic acid receptors (RAR), which hetero-dimerize with retinoic X receptors (RXR), forming active transcription factors. Cxcl13 is induced in stromal cells and the first LTi cells are attracted to this location, forming an initial cluster. This clustering results in the induction of cell surface expression of lymphotoxin-$\alpha_1\beta_2$ ($LT\alpha_1\beta_2$) on LTi cells (Cupedo et al. 2004a; Vondenhoff et al. 2009a). Expression of LT$\alpha_1\beta_2$ is needed to communicate with stromal cells, which express lymphotoxin receptor-β (LTβ-R) to further differentiate them to LTo cells. Signaling through LTβ-R leads to the expression of adhesion molecules, chemokines, as well as cytokines by these stromal cells (Cupedo et al. 2004a; Vondenhoff et al. 2009a). These molecules are required to retain the first cluster of LTi cells in place, attract more hematopoietic cells, and provide survival signals, respectively. In addition to the induction of these factors, LTβ-R signaling also leads to the expression of the lymph angiogenic factor VEGF-C, potentially involved in the attraction of lymphatic endothelial cells (Mounzer et al. 2010; Vondenhoff et al. 2009a). If signaling through LTβ-R does not take place, the initial cluster of LTi cells disappears and lymph nodes are not formed (Rennert et al. 1998; Vondenhoff et al. 2009a). Thus, signaling through LTβ-R is crucial for the definitive formation of the lymph nodes.

7.3 Lymphatic Endothelial Cell Specification

Lymphatic vasculature formation starts around embryonic day 10 post-coitus (E10) in mouse embryos with reprogramming of cardinal vein blood endothelial cells into lymphatic endothelial cells (LEC). In a polarized manner, a subset of blood endothelial cells in the cardinal vein start to express Sox18 and together with Coup-TFII (François et al. 2008; Srinivasan et al. 2010; Tammela and Alitalo 2010), these transcription factors subsequently initiate expression of transcription factor Prox-1, a hallmark for all lymphatic endothelial cells ((Wigle and Oliver 1999) see also Chap. 2 of this issue). After reprogramming, LECs bud off from the cardinal vein as strings or balloons (François et al. 2012; Hägerling et al. 2013) attracted to VEGF-C ((Karkkainen et al. 2004) and Chap. 13 of this issue) (Fig. 7.1). Subsequently, in a process called lymphangiogenesis, superficial lymph vessels sprout. However, it is not known what regulates the first events leading to the reprogramming of the blood vessel endothelial cells and why this process takes place at specific locations in the cardinal vein.

A key molecule shared between lymph node development and lymphatic vasculature formation is retinoic acid. Retinoic acid in combination with cAMP regulates LEC differentiation and size of the JLS (Marino et al. 2011). During lymph node formation, retinoic acid triggers Cxcl13 expression in LTo cells (van de Pavert et al. 2009) (Fig. 7.2). It has not been established whether retinoic acid is produced within LTo's or LECs themselves in an autocrine fashion, or whether it is provided by opposing cells, thus acting in a paracrine fashion. Paracrine action of retinoic acid has been reported to occur during embryonic development (Niederreither and Dollé 2008). It was shown that during embryogenesis, retinoic acid derived from nerve fibers, especially motor neurons, can mediate the differentiation of nearby cells (Berggren et al. 2001; Ji et al. 2006; Sockanathan and Jessell 1998). Since nerve fibers proximal to LN anlagen (van de Pavert et al. 2009) express high levels of the retinoic acid synthesizing enzyme Raldh2 and stimulating

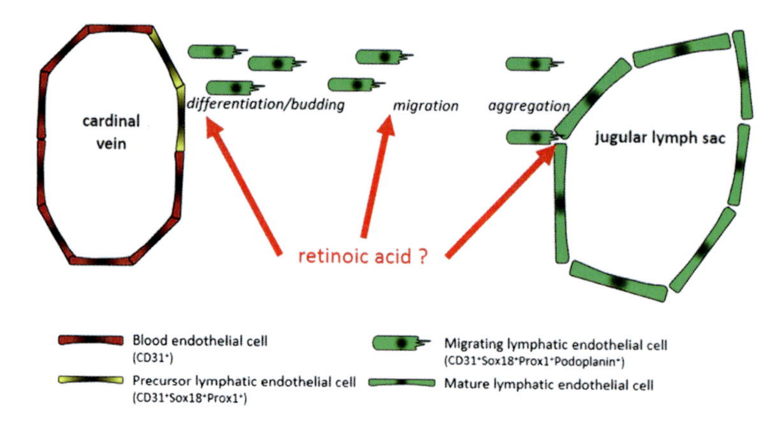

Fig. 7.1 Schematic overview of the events leading to differentiation of lymphatic endothelial cells and the processes that could possibly be influenced by retinoic acid, indicated by the *arrows*

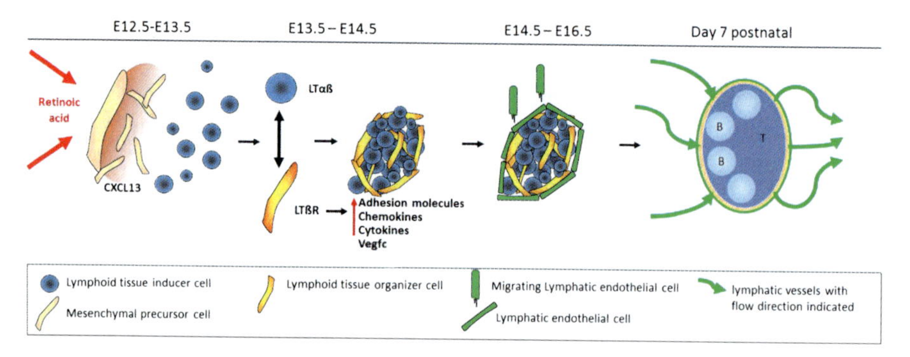

Fig. 7.2 Overview of the processes leading to attachment of the lymphatic endothelial cells to the lymph node anlagen in mice. Around the same time as the lymph sacs are formed, retinoic acid induces mesenchymal precursor cells to differentiate and synthesize chemokine Cxcl13 at specific locations. This leads to the attraction of the first hematopoietic lymphoid inducer cells and after initial clustering, interaction with the stromal organizer cells via induction of the lymphotoxin pathway. This results in amplification of chemokine synthesis and adhesion molecules to attract and retain more lymphoid tissue inducer cells. Also, stromal organizer cells now produce Vegfc, which can potently attract migrating lymphatic endothelial cells. These cells produce Ccl21, which further facilitates migration of lymphoid tissue inducer cells towards the lymph node anlagen. After birth, the lymph node is organized in B-cell and T-cell areas with connections to afferent lymphatics via the subcapsular sinus and efferent lymphatics

nerves resulted in upregulation of Cxcl13 in a retinoic acid receptor dependent manner, we proposed that nerve fibers release retinoic acid at specific locations. However, the mechanism by which retinoic acid affects specific differentiation stages towards LEC formation is unknown.

Since lymph node initiation and LEC differentiation both require retinoic acid, and since these structures develop within the same time frame, the source of retinoic acid could be the same. Noticeably, jugular lymph sacs are located in close proximity to cranial nerves in the neck region, which run parallel to the blood vessels. Remarkably, in a mouse model for Down syndrome, trisomy 16 embryos develop enlarged jugular lymph sacs and have differentially arranged cranial nerves IX, X (vagal nerve), and XI, when compared to wild-type controls (Bekker et al. 2005, 2006). Therefore, these data support the notion that nerves could provide retinoic acid at specific locations. This retinoic acid may in turn influence the endothelial cells of the cardinal vein and/or pre-LECs of the jugular lymph sacs to contribute to lymphatic endothelial formation and may additionally induce the initial production of Cxcl13 at locations where lymph nodes will develop.

7.4 Lymph Nodes and Lymphatics Develop in the Same Time Frame and Coordinated Fashion

The spatio-temporal similarities of lymph node and lymph sac development led Sabin (1909) to propose that lymph nodes originate from the primitive lymph sacs. However, studying mouse embryos which lacked lymphatic vasculature, by elimination of Prox-1, revealed normal initiation of lymph node development (Vondenhoff et al. 2009b), thereby disproving Sabin's original assumption. Thus, without the aid of lymphatic vasculature, LTi cells initially cluster at defined locations to start lymph node formation. During this initial phase, lymphatic vessels were not observed near or surrounding lymph node anlagen. While Cxcl13 was shown to be expressed within all developing peripheral lymph nodes, expression of Ccl21 was not always present (van de Pavert et al. 2009). Cxcl13 is essential for the initiation of peripheral lymph node formation, while the Ccr7/Ccl21 receptor-ligand pair is additive to the attractive function of Cxcl13-Cxcr5. This additive effect only becomes apparent in double-deficient mice, in which the Ccr7-Ccl19/21 as well as the Cxcr5/Cxcl13 ligand receptor pairs are disrupted (Luther et al. 2003). Thus, in Cxcl13- and Cxcr5-deficient mice, various peripheral lymph nodes fail to develop, while cervical and facial lymph node development still takes place. Only in the additional absence of Ccr7 in Cxcr5$^{-/-}$ mice, facial and cervical lymph nodes can no longer form, indicating the involvement of Ccr7 ligands in attracting LTi cells to these locations. Remarkably, when *Cxcl13*$^{-/-}$ mice were crossed to the *plt/plt* mutant mice, deficient for Ccl21 and Ccl19 expression in lymph nodes while Ccl21 is still expressed by LECs, facial and cervical lymph nodes could still form (Luther et al. 2003). This finding indeed disclosed a role for Ccl21 derived from LECs in attracting LTi cells towards the facial and cervical LNs, which are near the JLS. Therefore, ligands for Ccr7 only add to the attractive function of Cxcl13 during the initial cluster formation of LTi cells, since in Ccr7 single-deficient mice or *plt/plt* mice in which the Cxcl13/Cxcr5 axis is still intact, all lymph nodes are able to form in a normal fashion.

Also, in embryos that lacked synthesis of the retinoic acid synthesizing enzyme Raldh2 and consequently lacked Cxcl13 expression, clusters of LTi cells were observed around Ccl21-expressing cells in the cervical area (van de Pavert et al. 2009). However, the clusters in the *Raldh2*$^{-/-}$ embryos did not resemble normal-appearing lymph node anlagen, since Ccl21 is normally expressed at the rim of lymph node anlagen, while in the *Raldh2*$^{-/-}$embryos Ccl21 was expressed within these clusters of LTi cells. In conclusion, Ccl21 expressed by LECs can attract LTi cells, thereby contributing to lymph node formation, while it is not crucial for the early clustering of LTi cells.

After the initial process of LTi and LTo clustering until E13.5, LECs are observed adjacent to LN anlagen. At E14.5 until E16.5, depending on the location, lymph node anlagen are surrounded by a ring of lymphatic endothelial cells (Bénézech et al. 2010; Cupedo et al. 2004a; van de Pavert et al. 2009; Vondenhoff et al. 2009a, b) (Fig. 7.2). Ccl21 expression is very prominent in these cells, although it is not clear whether the lymphatic vasculature is able to facilitate transport of LTi cells at this embryonic stage. During early lymph node development, we observed that Cxcl13 and Ccl21 proteins were not expressed within the same (stromal) cells, although it was shown at the population level that LTo cells can express mRNA for both chemokines (Bénézech et al. 2010; Cupedo et al. 2004a; Honda et al. 2001; Vondenhoff et al. 2009a). Whether indeed LTo cells can make both Ccl21 and Cxcl13 or whether this is simply due to the fact that the population of LTo cells contains Ccl21-producing LECs and Cxcxl13-producing LTo cells will need further studies.

Since LN anlagen are initially formed by clusters of LTi cells, it is being assumed that lymphatic vessels grow towards LN anlagen (Fig. 7.2). This would suggest that LN anlagen secrete attracting factors. Indeed, we have shown that upon LTβR signaling, embryonic stromal cells respond by increased VEGF-C expression (Vondenhoff et al. 2009a), which is a potent LEC attraction molecule (Karkkainen et al. 2004; Tammela et al. 2007). Starting after E13.5 we observed that Lyve-1 positive cells were surrounding LN anlagen, which fits within the time frame in which LTβR signaling on LTo cells is initiated, resulting in the secretion of VEGF-C (Cupedo et al. 2004a; van de Pavert et al. 2009; Vondenhoff et al. 2009a). The time frame, in which the first VEGF-C is being produced within the developing lymph nodes, could be further visualized using a VEGF-C reporter embryo (Karkkainen et al. 2004). Such an analysis will allow better insight into the coordinated development of lymph nodes, LTβR signaling, and lymphatic vessel ingrowth to these structures. Summarizing, the first interactions of LTi cells with LTo cells by lymphotoxin-mediated signaling result in the induction of VEGF-C expression by LTo cells, which mediates the attraction of LECs towards the LN anlagen. The lymphatic cells that surround the lymph node will express Ccl21, which in turn can attract more LTi cells.

7.5 Interdependence of Lymph Nodes and Lymphatics

Without the lymphatic vasculature, as was shown in the *Prox1* KO embryos, LN formation was initiated normally. However, conditional deletion of *Prox-1*, using *Tie-2*, allowed for the analysis of later stages of LN development in the absence of Prox-1. In these embryos, lymphangiogenesis is severely compromised (Srinivasan et al. 2007) and only few scattered superficial LECs are present. In these conditional knockout embryos, LTi cluster size within developing lymph nodes at E17.5 was greatly reduced, coinciding with decreased numbers of LTo cells (Vondenhoff et al. 2009b). LECs are therefore necessary for LN development as they contribute to the further increase of LTi numbers, after the initial clustering of LTi resulting in normal-sized clusters of LTi and LTo cells.

Conversely, lymphatic vasculature formation can occur without the presence of lymph nodes as various mouse models devoid of lymph nodes, such as $RORc^{-/-}$ or the $LT\alpha^{-/-}$, contain lymphatic vessels (Sun et al. 2000; De Togni et al. 1994). However, it was shown that lymphotoxin-α-mediated signaling was contributing to lymphangiogenesis seen during inflammation (Mounzer et al. 2010).

Proper drainage through lymphatics requires connections to lymph nodes, and vice versa, lymphatics are necessary for proper functioning of lymph nodes. It was shown that interruption of afferent lymph vessels resulted in loss of markers and degeneration of high endothelial venules within the lymph node (Mebius et al. 1991, 1993) thereby severely compromising immune responses. Therefore, it is likely that lymph transport, and factors therein, is necessary for maintenance of the high endothelial venules and thus for normal functioning of lymph nodes.

And thus, while lymphatics are necessary for the lymph nodes to work properly, lymphatics also function best when lymph nodes are present. Disconnected lymphatics as a result of lymph node removal in cancer patients severely lead to disturbances in lymph drainage, which manifests as lymphedema. To allow for reattachment of the lymphatics and recovery of lymphedema, VEGF-C was shown to promote the establishment of functional lymphatics when provided together with lymph nodes upon transplantation in the axial region of lymph node excised mice (Tammela et al. 2007).

7.6 Neogenesis of Lymph Nodes and Lymph Vessels

Inflammation is a strong initiator of lymph node and lymphatic vessel remodeling. Lymph nodes expand by proliferation and differentiation of lymph node stromal cells, thus effectively increasing the scaffold necessary for immune interactions (Katakai et al. 2004). It was shown that B cells within lymph node B-cell follicles contribute to the expansion of the lymphatic vessels through their production of VEGF-A, thereby stimulating lymphangiogenesis (Angeli et al. 2006; Halin et al. 2007), while inhibition of lymphangiogenesis can be mediated by IFN-γ produced by T cells within lymph nodes ((Kataru et al. 2011), see also Chap. 9 of this issue).

Ectopic lymphoid organs can also be induced by placing a mesenteric lymph node suspension obtained from neonatal mice into the skin of mice (Cupedo et al. 2004b). In the recovered ectopic lymph nodes, the high endothelial venules and stromal populations in the B-cell follicle, the follicular dendritic cells, and the

T-cell region, i.e., the fibroblastic reticular cells, were of donor origin, while all hematopoietic cells were recipient derived. This implies that mesenchymal precursor cells, which were already exposed to LTβ-R-mediated signaling prenatally, can further differentiate into the proper lymph node stromal cell subsets and create the respective lymph node domains. Interestingly, also the LECs surrounding the induced ectopic lymph nodes were of donor origin and located at the expected location (Cupedo et al. 2004b).

Lymph node formation is a dynamic and flexible process depending on factors that were readily discovered by studying embryonic lymph node development. Analogous to embryonic formation, new lymphoid organs can be formed in adult during chronic inflammation, when ectopic lymph nodes (tertiary lymphoid organs, TLO) are formed (Aloisi and Pujol-Borrell 2006). Whether these organs are beneficial or detrimental to the host probably depends on the site and kind of inflammation, as TLOs are usually detrimental in autoimmune diseases such as rheumatoid arthritis (Drayton et al. 2006), while they may contribute to immune defense when combating pathogens (GeurtsvanKessel et al. 2009). The combination of neogenesis by placing precursor LTo cells together with the proper growth factors for LECs such as VEGF-C (Tammela et al. 2007) or CCBE1 (Bos et al. 2011) could result in a lymphoid organ with functional lymphatics and counteract lymphedema. Also, the use of retinoic acid to facilitate development of the lymphatic vasculature (Choi et al. 2012), while also inducing LN formation, will be an interesting topic to investigate. However, how exactly ectopic lymphoid organs are formed and which mesenchymal precursor cells are needed are currently unknown.

7.7 Conclusion

Initiation of LN formation and LEC differentiation are two processes, which occur independently. However, it is remarkable that these organs develop in the same time frame. Moreover, they share initiation cues such as retinoic acid. For the function of lymph nodes it is required that lymphatic vessels connect, as this will allow the influx of antigen-presenting cells and antigens, but intact lymphatic connections are also necessary for the maintenance of high endothelial venules, which are the entry site for naïve lymphocytes. Damage to the lymphatic vasculature will result in dysfunctional LN, and removal of lymph nodes in cancer treatment will often result in dysfunctional lymphatics.

Since the LN and lymphatic vasculature share retinoic acid as common developmental signal, it will be of interest to determine whether nerve fibers, adjacent to both developing lymph nodes as well as lymph sacs, are the source of retinoic acid. Further research is needed to precisely elucidate the role of nerves in the development of the LN and LEC, e.g., by use of conditional and cell-specific mutants. Moreover, for treatment of lymphedema, these common cues could be used to facilitate the formation of both lymphatic vasculature and lymphoid structures, in order to accelerate the recovery.

References

Aloisi, F., & Pujol-Borrell, R. (2006). Lymphoid neogenesis in chronic inflammatory diseases. *Nature Reviews Immunology, 6*, 205–217.

Angeli, V., Ginhoux, F., Llodrà, J., Quemeneur, L., Frenette, P. S., Skobe, M., et al. (2006). B cell-driven lymphangiogenesis in inflamed lymph nodes enhances dendritic cell mobilization. *Immunity, 24*, 203–215.

Ansel, K. M., Ngo, V. N., Hyman, P. L., Luther, S. A., Förster, R., Sedgwick, J. D., et al. (2000). A chemokine-driven positive feedback loop organizes lymphoid follicles. *Nature, 406*, 309–314.

Bekker, M. N., Arkesteijn, J. B., Van den Akker, N. M. S., Hoffman, S., Webb, S., Van Vugt, J. M. G., & Gittenberger-de Groot, A. C. (2005). Increased NCAM expression and vascular development in trisomy 16 mouse embryos: Relationship with nuchal translucency. *Pediatric Research, 58*, 1222–1227.

Bekker, M. N., Van den Akker, N. M. S., Bartelings, M. M., Arkesteijn, J. B., Fischer, S. G. L., Polman, J. A. E., et al. (2006). Nuchal edema and venous-lymphatic phenotype disturbance in human fetuses and mouse embryos with aneuploidy. *Journal of the Society for Gynecologic Investigation, 13*, 209–216.

Bénézech, C., White, A., Mader, E., Serre, K., Parnell, S., Pfeffer, K., et al. (2010). Ontogeny of stromal organizer cells during lymph node development. *Journal of Immunology, 184*, 4521–4530.

Berggren, K., Ezerman, E. B., McCaffery, P., & Forehand, C. J. (2001). Expression and regulation of the retinoic acid synthetic enzyme RALDH-2 in the embryonic chicken wing. *Developmental Dynamics, 222*, 1–16.

Bos, F. L., Caunt, M., Peterson-Maduro, J., Planas-Paz, L., Kowalski, J., Karpanen, T., et al. (2011). CCBE1 is essential for mammalian lymphatic vascular development and enhances the lymphangiogenic effect of vascular endothelial growth factor-C in vivo. *Circulation Research, 109*(5), 486–491.

Cherrier, M., Sawa, S., & Eberl, G. (2012). Notch, Id2, and RORγt sequentially orchestrate the fetal development of lymphoid tissue inducer cells. *Journal of Experimental Medicine, 209*, 729–740.

Choi, I., Lee, S., Chung, H. K., Lee, Y. S., Kim, K. E., Choi, D., et al. (2012). 9-cis retinoic acid promotes lymphangiogenesis and enhances lymphatic vessel regeneration: Therapeutic implications of 9-cis retinoic acid for secondary lymphedema. *Circulation, 125*(7), 872–882.

Cupedo, T., Jansen, W., Kraal, G., & Mebius, R. E. (2004a). Induction of secondary and tertiary lymphoid structures in the skin. *Immunity, 21*, 655–667.

Cupedo, T., & Mebius, R. E. (2005). Cellular interactions in lymph node development. *Journal of Immunology, 174*, 21–25.

Cupedo, T., Vondenhoff, M. F. R., Heeregrave, E. J., De Weerd, A. E., Jansen, W., Jackson, D. G., et al. (2004b). Presumptive lymph node organizers are differentially represented in developing mesenteric and peripheral nodes. *Journal of Immunology, 173*, 2968–2975.

De Togni, P., Goellner, J., Ruddle, N., Streeter, P., Fick, A., Mariathasan, S., et al. (1994). Abnormal development of peripheral lymphoid organs in mice deficient in lymphotoxin. *Science, 264*, 703–707.

Drayton, D. L., Liao, S., Mounzer, R. H., & Ruddle, N. H. (2006). Lymphoid organ development: From ontogeny to neogenesis. *Nature Immunology, 7*, 344–353.

François, M., Caprini, A., Hosking, B., Orsenigo, F., Wilhelm, D., Browne, C., et al. (2008). Sox18 induces development of the lymphatic vasculature in mice. *Nature, 456*, 643–647.

François, M., Short, K., Secker, G. A., Combes, A., Schwarz, Q., Davidson, T.-L., et al. (2012). Segmental territories along the cardinal veins generate lymph sacs via a ballooning mechanism during embryonic lymphangiogenesis in mice. *Developmental Biology, 364*, 89–98.

GeurtsvanKessel, C. H., Willart, M. A. M., Bergen, I. M., Van Rijt, L. S., Muskens, F., Elewaut, D., et al. (2009). Dendritic cells are crucial for maintenance of tertiary lymphoid structures in the lung of influenza virus-infected mice. *Journal of Experimental Medicine, 206*, 2339–2349.

Hägerling, R., Pollmann, C., Andreas, M., Schmidt, C., Nurmi, H., Adams, R. H., et al. (2013). A novel multistep mechanism for initial lymphangiogenesis in mouse embryos based on ultramicroscopy. *EMBO Journal, 32*(5), 629–644.

Halin, C., Tobler, N. E., Vigl, B., Brown, L. F., & Detmar, M. (2007). VEGF-A produced by chronically inflamed tissue induces lymphangiogenesis in draining lymph nodes. *Blood, 110*, 3158–3167.

Honda, K., Nakano, H., Yoshida, H., Nishikawa, S., Rennert, P., Ikuta, K., et al. (2001). Molecular basis for hematopoietic/mesenchymal interaction during initiation of Peyer's patch organogenesis. *Journal of Experimental Medicine, 193*, 621–630.

Ji, S.-J., Zhuang, B., Falco, C., Schneider, A., Schuster-Gossler, K., Gossler, A., & Sockanathan, S. (2006). Mesodermal and neuronal retinoids regulate the induction and maintenance of limb innervating spinal motor neurons. *Developmental Biology, 297*, 249–261.

Karkkainen, M. J., Haiko, P., Sainio, K., Partanen, J., Taipale, J., Petrova, T. V., et al. (2004). Vascular endothelial growth factor C is required for sprouting of the first lymphatic vessels from embryonic veins. *Nature Immunology, 5*, 74–80.

Katakai, T., Hara, T., Sugai, M., Gonda, H., & Shimizu, A. (2004). Lymph node fibroblastic reticular cells construct the stromal reticulum via contact with lymphocytes. *Journal of Experimental Medicine, 200*, 783–795.

Kataru, R. P., Kim, H., Jang, C., Choi, D. K., Koh, B. I., Kim, M., et al. (2011). T lymphocytes negatively regulate lymph node lymphatic vessel formation. *Immunity, 34*, 96–107.

Luther, S. A., Ansel, K. M., & Cyster, J. G. (2003). Overlapping roles of CXCL13, interleukin 7 receptor alpha, and CCR7 ligands in lymph node development. *Journal of Experimental Medicine, 197*, 1191–1198.

Marino, D., Dabouras, V., Brändli, A. W., & Detmar, M. (2011). A role for all-trans-retinoic acid in the early steps of lymphatic vasculature development. *Journal of Vascular Research, 48*, 236–251.

Mebius, R. E., Dowbenko, D., Williams, A., Fennie, C., Lasky, L. A., & Watson, S. R. (1993). Expression of GlyCAM-1, an endothelial ligand for L-selectin, is affected by afferent lymphatic flow. *Journal of Immunology, 151*, 6769–6776.

Mebius, R. E., Rennert, P., & Weissman, I. L. (1997). Developing lymph nodes collect CD4 + CD3- LTbeta + cells that can differentiate to APC, NK cells, and follicular cells but not T or B cells. *Immunity, 7*, 493–504.

Mebius, R. E., Streeter, P. R., Brevé, J., Duijvestijn, A. M., & Kraal, G. (1991). The influence of afferent lymphatic vessel interruption on vascular addressin expression. *Journal of Cell Biology, 115*, 85–95.

Mounzer, R. H., Svendsen, O. S., Baluk, P., Bergman, C. M., Padera, T. P., Wiig, H., et al. (2010). Lymphotoxin-alpha contributes to lymphangiogenesis. *Blood, 116*, 2173–2182.

Niederreither, K., & Dollé, P. (2008). Retinoic acid in development: Towards an integrated view. *Nature Reviews Genetics, 9*, 541–553.

Rennert, P. D., James, D., Mackay, F., Browning, J. L., & Hochman, P. S. (1998). Lymph node genesis is induced by signaling through the lymphotoxin beta receptor. *Immunity, 9*, 71–79.

Sockanathan, S., & Jessell, T. M. (1998). Motor neuron-derived retinoid signaling specifies the subtype identity of spinal motor neurons. *Cell, 94*, 503–514.

Srinivasan, R. S., Dillard, M. E., Lagutin, O. V., Lin, F.-J. J., Tsai, S., Tsai, M.-J. J., et al. (2007). Lineage tracing demonstrates the venous origin of the mammalian lymphatic vasculature. *Genes and Development, 21*, 2422–2432.

Srinivasan, R. S., Geng, X., Yang, Y., Wang, Y., Mukatira, S., Studer, M., et al. (2010). The nuclear hormone receptor Coup-TFII is required for the initiation and early maintenance of Prox1 expression in lymphatic endothelial cells. *Genes and Development, 24*, 696–707.

Sun, Z., Unutmaz, D., Zou, Y. R., Sunshine, M. J., Pierani, A., Brenner-Morton, S., et al. (2000). Requirement for RORgamma in thymocyte survival and lymphoid organ development. *Science, 288*, 2369–2373.

Tammela, T., & Alitalo, K. (2010). Lymphangiogenesis: Molecular mechanisms and future promise. *Cell, 140*, 460–476.

Tammela, T., Saaristo, A., Holopainen, T., Lyytikkä, J., Kotronen, A., Pitkonen, M., et al. (2007). Therapeutic differentiation and maturation of lymphatic vessels after lymph node dissection and transplantation. *Nature Medicine, 13*, 1458–1466.

Van de Pavert, S. A., & Mebius, R. E. (2010). New insights into the development of lymphoid tissues. *Nature Reviews Immunology, 10*, 664–674.

Van de Pavert, S. A., Olivier, B. J., Goverse, G., Vondenhoff, M. F., Greuter, M., Beke, P., et al. (2009). Chemokine CXCL13 is essential for lymph node initiation and is induced by retinoic acid and neuronal stimulation. *Nature Immunology, 10*, 1193–1199.

Vondenhoff, M. F., Greuter, M., Goverse, G., Elewaut, D., Dewint, P., Ware, C. F., et al. (2009a). LTbetaR signaling induces cytokine expression and up-regulates lymphangiogenic factors in lymph node anlagen. *Journal of Immunology, 182*, 5439–5445.

Vondenhoff, M. F., Van de Pavert, S. A., Dillard, M. E., Greuter, M., Goverse, G., Oliver, G., & Mebius, R. E. (2009b). Lymph sacs are not required for the initiation of lymph node formation. *Development, 136*, 29–34.

Wigle, J. T., & Oliver, G. (1999). Prox1 function is required for the development of the murine lymphatic system. *Cell, 98*, 769–778.

Chapter 8
Platelets in Lymph Vessel Development and Integrity

Steve P. Watson, Kate Lowe, and Brenda A. Finney

Abstract Blood platelets have recently been proposed to play a critical role in the development and repair of the lymphatic system. The platelet C-type lectin receptor CLEC-2 and its ligand, the transmembrane protein Podoplanin, which is expressed at high levels on lymphatic endothelial cells (LECs), are required to prevent mixing of the blood and lymphatic vasculatures during mid-gestation. A similar defect is seen in mice deficient in the tyrosine kinase Syk, which plays a vital role in mediating platelet activation by CLEC-2. Furthermore, blood-lymphatic mixing is also present in mice with platelet-/megakaryocyte-specific deletions of CLEC-2 and Syk, suggesting that the phenotype is platelet in origin. The molecular basis of this effect is not known, but it is independent of the major platelet receptors that support hemostasis, including integrin αIIbβ3 (GPIIb-IIIa). Radiation chimeric mice reconstituted with CLEC-2-deficient or Syk-deficient bone marrow exhibit blood-lymphatic mixing in the intestines, illustrating a role for platelets in repair and growth of the lymphatic system. In this review, we describe the events that led to the identification of this novel role of platelets and discuss possible molecular mechanisms and the physiological and pathophysiological significance.

8.1 Introduction

Blood platelets have recently been recognized to play a critical role in the development and repair of the lymphatic system. Activation of the platelet C-type lectin receptor CLEC-2 by the transmembrane protein Podoplanin on lymphatic endothelial cells (LECs) prevents mixing of the blood and the lymphatic vasculatures. This involves a pathway that is independent of the major platelet receptors that support hemostasis, including integrin αIIbβ3 (GPIIb-IIIa). Radiation chimeric mice

S.P. Watson (✉) • K. Lowe • B.A. Finney
Centre for Cardiovascular Sciences, The College of Medical and Dental Sciences, University of Birmingham, Birmingham B15 2TT, UK
e-mail: s.p.watson@bham.ac.uk

F. Kiefer and S. Schulte-Merker (eds.), *Developmental Aspects of the Lymphatic Vascular System*, Advances in Anatomy, Embryology and Cell Biology 214,
DOI 10.1007/978-3-7091-1646-3_8, © Springer-Verlag Wien 2014

reconstituted with CLEC-2-deficient bone marrow also exhibit blood-lymphatic mixing, illustrating a role in repair and growth of the lymphatic system. We describe the events that led to the identification of this novel role of platelets and discuss hypotheses on the underlying mechanism and the physiological and pathophysiological significance.

8.2 CLEC-2

The C-type lectin receptor CLEC-2 (gene name *CLEC-1b*) was first described as a transcript in a subpopulation of myeloid cells based on its sequence homology to other C-type lectins (Colonna et al. 2000). Several years later, CLEC-2 was identified in platelets by a combination of affinity chromatography using the snake venom toxin rhodocytin and proteomics (Suzuki-Inoue et al. 2006) and on megakaryocytes using serial analysis of gene expression technology (Senis et al. 2007). CLEC-2 was later shown to be expressed at a much lower level on a subpopulation of myeloid cells, including dendritic cells, neutrophils, and natural killer cells in mice, and to be upregulated on these and other hematopoietic lineages, but not on platelets, upon inflammatory challenge (Mourao-Sa et al. 2011).

CLEC-2 is a glycosylated type II transmembrane protein which contains an extracellular C-type lectin-like domain, which lacks the conserved amino acids for binding to sugars. CLEC-2 has a short cytoplasmic tail with a single conserved YxxL motif downstream of a triacidic amino acid sequence known as a hemi-immunoreceptor tyrosine-based activation motif (hemITAM). The closely related ITAM sequence has two YxxLs and forms the key signalling arm of a variety of Fc, antigen, and immunoglobulin receptors, including the platelet collagen receptor, the GPVI-FcR γ-chain complex (Watson et al. 2010). CLEC-2 stimulates powerful platelet activation through Src and Syk tyrosine kinases and a variety of adapter and effector proteins, including the cytosolic adapter, SLP-76, leading to activation of PLCγ2 (Watson et al. 2010) as shown in Fig. 8.1.

Mice with homozygous deletions in Syk, SLP-76, and PLCγ2 in the hemITAM signalling cascade have characteristic blood-filled lymphatic vessels in mid-gestation and a high level of lethality at the time of birth (Abtahian et al. 2003; Cheng et al. 1995; Ichise et al. 2009; Turner et al. 1995). A similar defect is seen in mice double deficient in Btk and Tec tyrosine kinases (unpublished). The extent of neonatal death varies from nearly all mice in the case of Syk to approximately one-third of mice deficient in SLP-76 and PLCγ2. The increased viability of the latter is likely to be due to a limited degree of platelet activation compared to a complete loss of platelet activation with Syk deficiency (Fig. 8.1). Death is believed to be caused by a failure to inflate the lungs as a result of defective lymphatic function and fluid retention (Finney et al. 2012). Significantly, these features are also found in mice deficient in CLEC-2.

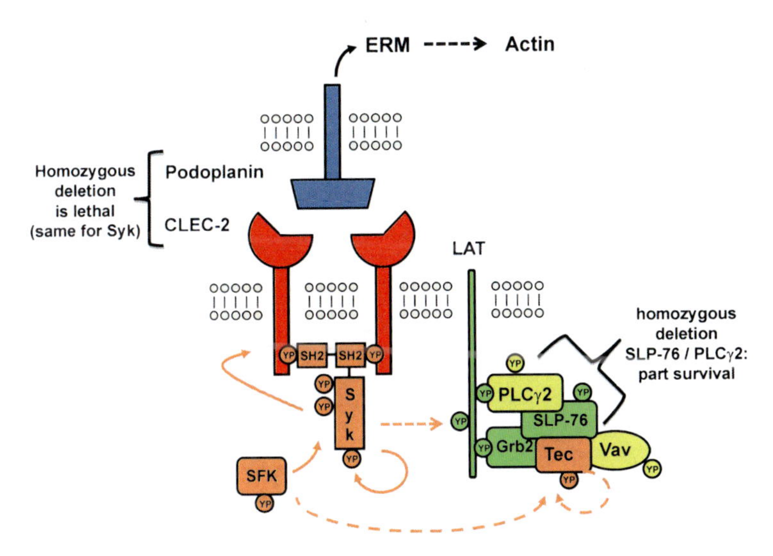

Fig. 8.1 Schematic of CLEC-2 signalling. CLEC-2 is expressed as a dimer on the surface of platelets. Binding of one Podoplanin to dimeric CLEC-2 leads to phosphorylation of Syk by a Src family kinase (SFK). In turn, Syk stimulates phosphorylation of the hemITAM in the CLEC-2 cytosolic chain. The binding of the tandem SH2 domains of Syk to two phosphorylated hemITAMs induces the formation of a LAT signalosome that involves a series of adapter proteins (*green*), including SLP-76, Gads, and Grb2 (depicted as Grb2 in the figure), and effector enzymes (*yellow*), including Vav and PLCγ2. Src, Syk, and Tec family kinases (*brown*) mediate phosphorylation of proteins in the LAT signalosome. Podoplanin signals via the ezrin, radixin, and moesin (ERM proteins) and small G proteins. Mice with constitutive deletions of Podoplanin, CLEC-2, Syk, SLP-76, and PLCγ2 have blood-filled lymphatics in mid-gestation. The increased survival of mice deficient in SLP-76 and PLCγ2 relative to mice deficient in Podoplanin, CLEC-2, and Syk is due to retention of limited signalling. By the same token, the absence of a phenotype in mice deficient in the other proteins in the signalling cascade is believed to be because of a sufficient level of activation of PLCγ2

8.3 Podoplanin Is the Endogenous Ligand for CLEC-2

Evidence for an endogenous ligand for CLEC-2 was provided by the demonstration of binding of the C-type lectin receptor to HIV-1 produced by human embryonic kidney (HEK) 293 T cells irrespective of the viral envelope protein (Chaipan et al. 2006). Podoplanin was later identified as the causative ligand (Christou et al. 2008). In an independent approach, the similar kinetics of platelet activation by tumor cell lines to that of rhodocytin led to identification of Podoplanin as an endogenous ligand for CLEC-2 (Suzuki-Inoue et al. 2007).

Podoplanin, also known as gp38, aggrus, and T1α, is a heavily sialylated transmembrane protein with a short cytoplasmic tail containing a binding site for the ezrin-radixin-moesin (ERM) family of actin-binding proteins. Podoplanin is restricted to mammals and has no homology to other proteins. Podoplanin is widely expressed and is found at high levels on LECs, type 1 lung alveolar cells, kidney podocytes, ciliary epithelium, and the choroid plexus (Astarita et al. 2012).

Podoplanin is also expressed on macrophages and Th17 T cells in response to inflammatory challenge (Kerrigan et al. 2012; Peters et al. 2011). Podoplanin is not expressed on blood endothelial cells.

The interaction of Podoplanin and CLEC-2 gives rise to reciprocal regulation of Src and Syk kinases by CLEC-2 and ERM proteins by Podoplanin (Fig. 8.1). CLEC-2 regulates powerful cell activation, whereas the effect of clustering of Podoplanin is cell-dependent, with increased phosphorylation of ERM proteins reported in some but not all studies (Martin-Villar et al. 2006). The few studies on Podoplanin signalling have primarily been performed in tumor cell lines, although Podoplanin has been shown to activate the small G proteins, CDC42 and RhoA, in human lung microvascular LECs (Navarro et al. 2011).

8.4 Mice Deficient in Podoplanin and CLEC-2 Have Blood-Filled Lymphatics

While it has been known since 2003 that mice deficient in Podoplanin have defective lymphatics and a normal blood vasculature (Schacht et al. 2003), it is only recently that they were shown to phenocopy mice deficient in Syk, SLP-76, PLCγ2, and CLEC-2. Constitutive deletions of any of these five proteins give rise to characteristic blood-filled lymphatics and edema in mid-gestation as typified by a *CLEC-1b*-deficient mouse (Fig. 8.2) and a high level of perinatal mortality (Bertozzi et al. 2010; Finney et al. 2012; Suzuki-Inoue et al. 2010; Uhrin et al. 2010). Mice which survive to suckle have chyloascites, reflecting impaired lymphatic function in the intestines (Abtahian et al. 2003; Finney et al. 2012). The high rate of mortality is associated with a failure to fully inflate the lungs, possibly because of fluid retention caused by the defect in lymphatic function (Finney et al. 2012; Schacht et al. 2003), although a role for Podoplanin on type 1 lung alveolar cells cannot be ruled out.

Additional defects in development have been described in Podoplanin- and CLEC-2-deficient mice, although a systematic comparison has yet to be reported. Hemorrhaging has been reported in the brain in mid-gestation in CLEC-2-deficient mice (Finney et al. 2012; Tang et al. 2010), while the absence of lymph nodes (Peters et al. 2011) and developmental defects in the proepicardial organ of the heart (Mahtab et al. 2008) have been described in Podoplanin-deficient mice.

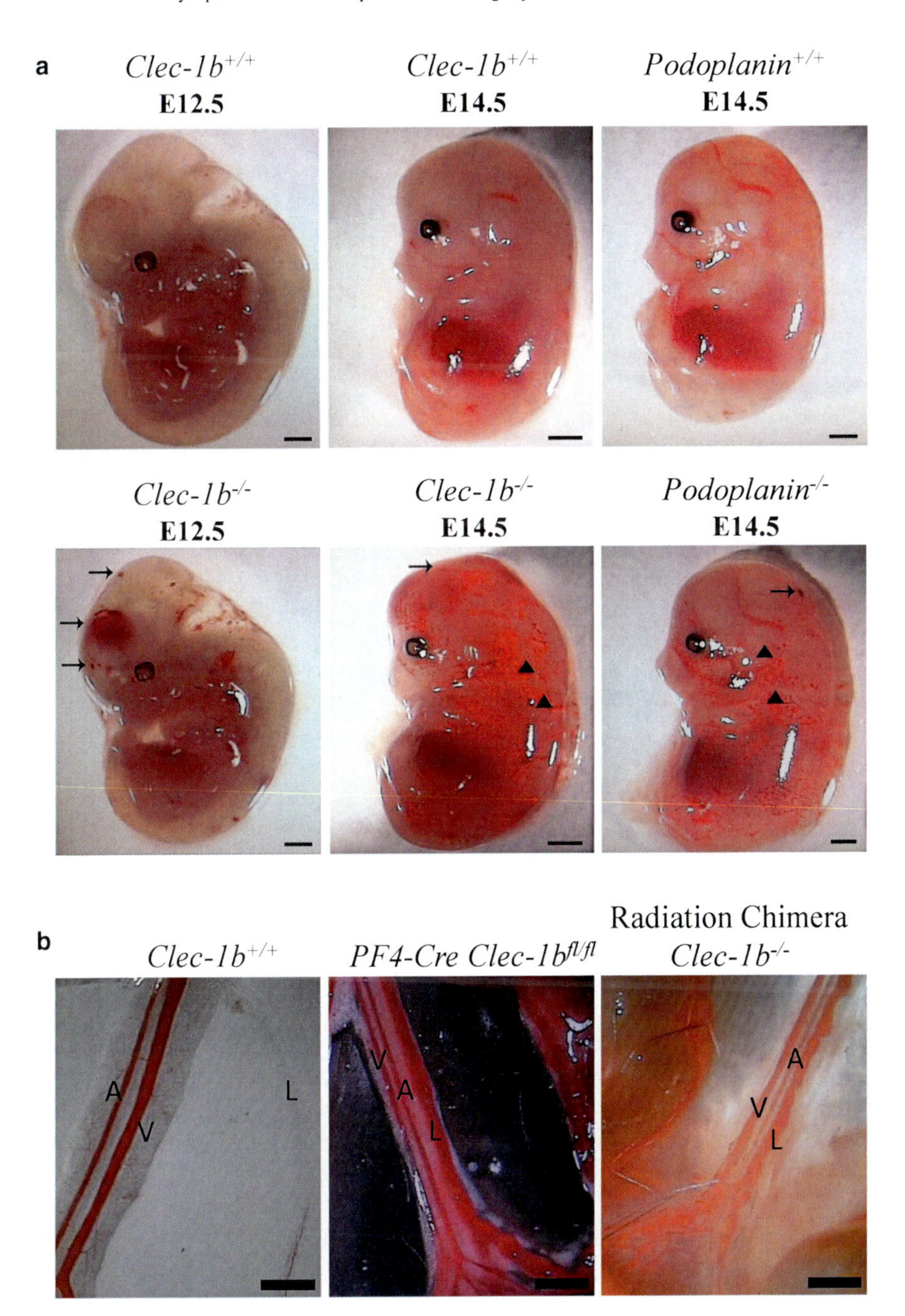

Fig. 8.2 (**a**) CLEC-2 deficiency results in defective lymphatic development and hemorrhaging in the brain. At E12.5, CLEC-2 deficient mice (*Clec-1b*$^{-/-}$) manifest with severe brain hemorrhages (*black arrows*). At E14.5, brain hemorrhages are still visible with the addition of blood-filled lymphatics in the skin (*black arrowheads*). The same phenotype is seen in mice lacking Podoplanin at E14.5 (*Podoplanin*$^{-/-}$) but not in control littermates. The figures shown are

8.5 The Defect in Lymphatic Development Is Platelet Mediated

The blood-filled vessels on the surface of Syk- and SLP-76-deficient embryos were identified as lymphatic vessels (Abtahian et al. 2003), but the blood-lymphatic mixing was attributed to a defect in circulating endothelial progenitor cells using lineage-mapping studies (Sebzda et al. 2006). This explanation was not supported by subsequent lineage-mapping studies using Cre-recombinase under the control of the endogenous Syk promoter and crossing to *R26R-lacZ* reporter mice, as this failed to detect expression of Syk at any developmental stage in the vascular endothelium (Böhmer et al. 2010).

It was not until Podoplanin was identified as the endogenous ligand for CLEC-2 that attention began to focus on the role of CLEC-2 and platelets. The demonstration of blood-filled lymphatics in mid-gestation in Podoplanin-deficient mice and in mice treated with the antiplatelet agent aspirin provided indirect evidence for platelet involvement (Uhrin et al. 2010). This was followed by the description of blood-lymphatic mixing at the same stage in development in mice deficient in the transcription factor, Meis1, which lack functional megakaryocytes, or following targeted destruction of the megakaryocyte/platelet lineage using diphtheria toxin (Carramolino et al. 2010). Blood-filled lymphatics in mid-gestation were subsequently observed in mice with a constitutive or megakaryocyte-/platelet-specific ablation of CLEC-2, Syk, or SLP-76 (Bertozzi et al. 2010; Finney et al. 2012; Suzuki-Inoue et al. 2010). Megakaryocyte-/platelet-specific targeting was achieved in all of these studies using a PF4-Cre transgene. While this approach does not distinguish between a role for platelets or megakaryocytes, the presence of blood-lymphatic mixing in the intestines of radiation chimeric mice reconstituted with CLEC-2 or Syk-deficient bone marrow favors a platelet-dependent process due to the low numbers of circulating megakaryocytes (Bertozzi et al. 2010; Finney et al. 2012; Hughes et al. 2010; Ichise et al. 2009). In contrast to the constitutive knockouts, *PF4-Cre.CLEC-1b*$^{fl/fl}$ and *PF4-Cre.Syk*$^{fl/fl}$ mice are viable and survive for upwards of 50 weeks, by which time their gross appearance is similar to their littermates.

Although the PF4-Cre transgene is partially expressed in other hematopoietic lineages (Calaminus et al. 2012), the use of various transgenic and Cre-expressing mice provides powerful evidence against the involvement of an additional lineage in the blood-lymphatic mixing. In an early study, Sebzda et al. (2006) used a GATA1-GFPSLP-76 transgene, which drives SLP-76 expression in platelets, erythrocytes, and eosinophils, to rescue the blood-lymphatic mixing in SLP-76-deficient

---

Fig. 8.2 (continued) representative of more than six embryos of each time-point and genotype. Scale bars represent 1 mm. (**b**) Blood-filled lymphatics in the mesentery. Blood-filled lymphatic vessels (L) are seen in the mesentery next to the arteries (A) and veins (V) as a result of developmental defects (*PF4-Cre Clec-1b*$^{fl/fl}$) or defective repair mechanisms (radiation chimera *Clec-1b*$^{-/-}$)

mice. Subsequently, we and others have shown that deletion of CLEC-2, Syk, or SLP-76 in other lineages, including blood endothelial cells, macrophages, and dendritic cells, using lineage-specific Cre-expression does not induce blood-lymphatic mixing (Bertozzi et al. 2010; Finney et al. 2012). Furthermore, deletion of SLP-76 or Syk using a Vav1.Cre transgene, which drives efficient recombination in all hematopoietic lineages, induced an almost identical phenotype to that of PF4-Cre, including viability at birth (Bertozzi et al. 2010; Finney et al. 2012).

The evidence presented above provides compelling evidence that blood-lymphatic mixing results from loss of Podoplanin-mediated activation of CLEC-2 and Syk in platelets. Nevertheless, the observation that the PF4-transgene induces expression of Cre in a subpopulation of other hematopoietic lineages and that CLEC-2 is expressed at low level on other hematopoietic cells leaves open the possibility that this phenotype could involve another hematopoietic lineage. In this context, Böhmer and colleagues have identified a population of Syk-expressing M2 myeloid cells that induce lymphangiogenesis in a dorsal skin chamber assay through a pathway that is dependent on secretion of VEGF-C and/or -D (Böhmer et al. 2010).

8.6 The Molecular Basis of the Role of Platelets in Lymphatic Development

The initial proposal that the blood-lymphatic mixing is due to the loss of CLEC-2-driven thrombus formation at the site of separation of the lymphatic and venous systems is no longer thought to be correct. Studies using high-resolution ultramicroscopy and electron microscopy have failed to find evidence for a connection between lymphatic and blood vasculatures at the time that Prox1-expressing endothelial cells migrate away from the cardinal vein ((Hagerling et al. 2013; Yang et al. 2012) see also Chaps. 2 and 13 of this issue). Furthermore, a role for thrombus formation in development of the lymphatic system seems unlikely as blood-lymphatic mixing is absent in mice deficient in the major platelet receptors that support hemostasis, including integrin αIIbβ3 and GPIb-IX-V. Thus, there is no requirement for CLEC-2-dependent thrombus formation at the initiation of lymphatic development.

Platelets are powerful secretory cells, releasing the contents of dense, α-, and lysosomal granules, as well as a variety of lipid mediators including sphingosine 1-phosphate (S1-P), upon activation. They also support the generation of lysophosphatidic acid (LPA) by the extracellular lipase autotaxin and form a procoagulant surface for thrombin generation. Any one of these pathways could support a role for CLEC-2 in lymphatic development, although several can be ruled out by the absence of blood-lymphatic mixing during gestation in relevant mouse models.

Dense granules are the source of several feedback mediators that support hemostasis, including ADP. The absence of dense granules gives rise to a characteristic syndrome known as Hermansky-Pudlak syndrome, which is associated with albinism. Platelet α-granules contain over a hundred proteins, which support a variety of functions, including hemostasis, angiogenesis, and chemoattraction. Platelet α-granules are absent in grey platelet syndrome and ARC syndrome caused by recessive mutations in Nbeal2 or the trafficking proteins VPS33b or VIPAR, respectively. Blood-lymphatic mixing has not been described in dense or α-granule secretion disorders in either human or mouse models, including mice deficient in Nbeal2 (Deppermann et al. 2013). This absence of blood-lymphatic mixing in Nbeal2-deficient mice is not consistent with the proposal that this phenotype is mediated by release of BMP-9 from platelet α-granules (Osada et al. 2012).

Scott syndrome is a rare platelet bleeding disorder that is associated with impaired expression of procoagulant phospholipids on the platelet surface, including phosphatidylserine. The causative gene in Scott syndrome is the transmembrane Ca^{2+} channel, TMEM16F. Although lymphatic defects are not observed in Scott syndrome patients, a role for procoagulant activity cannot be excluded as other pathways support phosphatidylserine exposure. A role for lysosomal secretion, or the biologically active lipids S1-P or LPA, which have powerful effects on the migration of blood endothelial cells, is also possible. Mice deficient in the extracellular lipase autotaxin die prior to development of the lymphatic system and so provide no insight into the potential role of LPA in this process.

The prevention of blood-lymphatic mixing could also involve regulation of Podoplanin signalling by CLEC-2. If this is the case, CLEC-2-dependent platelet activation must be required either to stabilize adhesion to LECs through clustering of CLEC-2 or regulation of a platelet receptor with a counter ligand on the LEC surface. Podoplanin regulates migration of cancer cells through actin polymerization and cell adhesion (Astarita et al. 2012). Cross-linking of Podoplanin using a specific antibody or recombinant CLEC-2 has been reported to inhibit migration of LECs in a transwell assay (Finney et al. 2012; Osada et al. 2012). Platelets are also inhibitory in this assay with partial recovery seen in the absence of Syk or CLEC-2 (Finney et al. 2012). Platelets also prevent formation of lymphatic tubes on a matrigel surface through a pathway that is partially dependent on CLEC-2 or Syk (Finney et al. 2012). It is unclear if this is due to disruption of LEC-LEC interaction, altered migration, or lymphatic precursor proliferation. It is unclear if these effects contribute to blood-lymphatic mixing in vivo.

The interaction of CLEC-2 and Podoplanin could be required for binding of a second platelet receptor to a counterpart ligand on the LEC surface. Platelet integrins are regulated by outside-in signalling but are unlikely to play a role in the blood-lymphatic mixing as mice deficient in the global regulators of integrins in platelets, talin, do not have a defect in lymphatic development. Ephrins, Eph kinases, and semaphorins are expressed on both platelets and LECs and are widely recognized for their role in migration. Several platelet receptors are expressed exclusively on intracellular secretory granules and expression of one or more of these on the surface of an activated platelet could account for the requirement for platelet activation.

8.7 Where Is the Connection Between the Lymphatic and Blood Vasculatures?

The development of the lymphatic system is described in Chaps. 2, 7, and 13 of this issue. In brief, lymphatic development begins at E9.5 when Prox1+ LECs migrate away from the cardinal vein and superficial venous plexus as streams of attached single cells form the first lumenized, lymphatic structures which are known collectively as the lymphatic sacs, but which are now recognized to be two distinct structures, the dorsal peripheral longitudinal lymphatic vessel (PLLV) and primordial thoracic duct (pTD) (Hagerling et al. 2013; Yang et al. 2012). Both large lumenized lymphatic vessels can be observed from E11.0 coincident with expression of Podoplanin (Hagerling et al. 2013). Superficial lymphatic vessels sprouting from the dorsal PLLV develop into the dermal lymphatic vessels, with the lymphatic vasculature appearing in the skin around E12–E15 and in the intestine from E18.

The presence of blood in the lymphatic sacs of Syk- and SLP-76-deficient mice is seen as early as E11.5 (Abtahian et al. 2003), and from E12.0 onwards, densely packed red blood cells can be seen in the primordial thoracic duct adjacent to the cardinal vein (Hagerling et al. 2013). However, it is not until E14.5, when the cutaneous lymphatics have begun to develop, that prominent blood-filled vessels can be seen on the surface of embryos (Böhmer et al. 2010). Marked edema can also be seen at this stage. Blood-lymphatic mixing is seen in the mesentery in the intestine coincident with growth of the lymphatics. The presence of blood in the cutaneous vessels, and the associated edema, begins to resolve towards the end of embryogenesis, although the extent of resolution varies according to the nature of the gene "knockout." In the case of SLP-76-deficient neonates, the connection between the blood and cutaneous lymphatics is lost prior to birth (Chen et al. 2012) and this may contribute to their increased survival rate relative to mice with constitutive deficiencies in Podoplanin, CLEC-2, and Syk. Mice with a PF4-Cre-targeted deficiency in CLEC-2 or Syk exhibit variable severity of edema at birth, with some mice displaying edema in one or all of their extremities and others having no outwardly visible symptoms.

Blood-lymphatic mixing is present in the mesenteric lymphatics of all viable mice, which show blood-lymphatic mixing in mid-gestation, as illustrated in a *PF4-Cre, CLEC-1b$^{fl/fl}$* mouse (Fig. 8.2b). The presence of large, dilated, blood-filled collecting vessels that run alongside a vein and artery can be readily seen, along with smaller blood-filled lymphatics. Uninterrupted flow is present in the large vessels, which are continuous with the venous system as shown by FITC-dextran injections into the arterial circulation and the almost simultaneous labelling of the venous and lymphatic vasculatures (Abtahian et al. 2003; Chen et al. 2012; Finney et al. 2012). The vessels have been remodelled by the continuous flow and have lost LEC markers and now express those of blood endothelial cells (Chen et al. 2012). The smaller, blood-filled vessels label poorly with the FITC-dextran and retain the properties of LECs, suggesting that their connections are poorly developed or that they have resolved.

In their original study, Abtahian et al. (2003) reported the presence of chimeric vessels in the neck and chest that could be partially stained with LYVE-1. Others have also reported the presence of vessels that can be co-stained with lymphatic and blood endothelial markers (Ichise et al. 2009; Suzuki-Inoue et al. 2010). As yet, however, there has been no systematic description of the frequency, significance, and anatomical location of these connections, and it is likely that this will require the use of a 3-D reconstitution approach to establish an overall picture such as that used to establish the events that underlie the initiation of lymphatic development (Hagerling et al. 2013).

8.8 The Role of Platelets in the Maintenance and Repair of the Lymphatic System

It is widely recognized that platelets are normally absent in the lymphatic system, suggesting that they play no role in its day-to-day maintenance. Furthermore, mice with PF4-Cre-targeted deletions of CLEC-2 or Syk are viable and show no evidence of increased edema with aging up to 50 weeks relative to controls, despite the presence of blood-filled lymphatics and edema in mid-gestation (Finney et al. 2012). On the other hand, leakage of blood into the mesenteric lymphatics has been reported after six months in a tamoxifen-inducible T-synthase mouse which lacks functional Podoplanin on LECs (Fu et al. 2008). It is not yet known whether this is mediated by an interaction with CLEC-2 on platelets.

The expression of CLEC-2 on platelets is critical for mediating repair and growth of the lymphatic system following radiation exposure. Mice subjected to a dose of radiation that induces bone marrow ablation and extensive destruction of the intestinal microvasculature show significant signs of malnutrition and have to be euthanized within 3–6 months following reconstitution with CLEC-2-, Syk-, SLP-76-, or PLCγ2-deficient bone marrow (Abtahian et al. 2003; Finney et al. 2012; Ichise et al. 2009; Kiefer et al. 1998). Dissection of the mice at this time reveals extensive mixing of the blood and lymphatic systems in the intestines, which is similar in appearance to that in the constitutive and PF4-Cre-specific models (Fig. 8.2). In contrast, mice reconstituted with fetal liver from littermate controls are healthy.

It is not known whether platelets are required for repair or formation of the lymphatics in response to other forms of injury. It is also unclear whether platelets are required for growth of new vessels in various pathologies, including solid tumors, which need to generate their own blood and lymphatic systems. In such a scenario however, it can be speculated that pharmacological targeting of CLEC-2 or Podoplanin may prevent both tumor growth and Podoplanin-mediated metastasis (Astarita et al. 2012).

8.9 Concluding Remarks

Recent developments have highlighted a novel role for platelets in the development and repair of the lymphatic system through the interaction of Podoplanin and CLEC-2. The molecular basis of this is to be established, but it appears to be unrelated to the role of platelets in hemostasis or secretion of platelet dense and α-granules. Platelets are also required for the repair of the lymphatic system following radiation, and it will be important to establish whether they participate in the growth of new lymphatics in other forms of injury and pathologies, including cancer.

Podoplanin and CLEC-2 are implicated in several other pathways in development and in response to injury and inflammation. This is illustrated by the critical role of CLEC-2 in the development of the vasculature in the brain in mid-gestation (Finney et al. 2012) and maintenance of vascular integrity in thrombocytopenic mice subject to inflammatory challenge (Boulaftali et al. 2013). CLEC-2 and Podoplanin are both upregulated on subpopulations of hematopoietic cells in the presence of LPS. These observations suggest a much wider role for the Podoplanin-CLEC-2 axis and it will be important to establish to what extent this role is dependent on platelet activation and hemostasis.

Acknowledgements The research in the authors' laboratory is supported by the Wellcome Trust, British Heart Foundation, and Medical Research Council. We are grateful to Drs Alexander Brill, Craig Hughes, Leyre Navarro-Nunez, and Alice Pollitt for their constructive comments on the manuscript.

References

Abtahian, F., Guerriero, A., Sebzda, E., Lu, M. M., Zhou, R., Mocsai, A., et al. (2003). Regulation of blood and lymphatic vascular separation by signaling proteins SLP-76 and Syk. *Science, 299*, 247–251.

Astarita, J. L., Acton, S. E., & Turley, S. J. (2012). Podoplanin: Emerging functions in development, the immune system, and cancer. *Frontiers in Immunology, 3*, 283.

Bertozzi, C. C., Schmaier, A. A., Mericko, P., Hess, P. R., Zou, Z., Chen, M., et al. (2010). Platelets regulate lymphatic vascular development through CLEC-2-SLP-76 signaling. *Blood, 116*, 661–670.

Böhmer, R., Neuhaus, B., Bühren, S., Zhang, D., Stehling, M., Böck, B., et al. (2010). Regulation of developmental lymphangiogenesis by Syk + leukocytes. *Developmental Cell, 18*, 437–449.

Boulaftali, Y., Hess, P. R., Getz, T. M., Cholka, A., Stolla, M., Mackman, N., et al. (2013). Platelet ITAM signaling is critical for vascular integrity in inflammation. *Journal of Clinical Investigation, 123*(2), 908–916.

Calaminus, S. D., Guitart, A., Sinclair, A., Schachtner, H., Watson, S. P., Holyoake, T. L., et al. (2012). Lineage tracing of Pf4-Cre marks hematopoietic stem cells and their progeny. *PLoS One, 7*, e51361.

Carramolino, L., Fuentes, J., Garcia-Andres, C., Azcoitia, V., Riethmacher, D., & Torres, M. (2010). Platelets play an essential role in separating the blood and lymphatic vasculatures during embryonic angiogenesis. *Circulation Research, 106*, 1197–1201.

Chaipan, C., Soilleux, E. J., Simpson, P., Hofmann, H., Gramberg, T., Marzi, A., et al. (2006). DC-SIGN and CLEC-2 mediate human immunodeficiency virus type 1 capture by platelets. *Journal of Virology, 80*, 8951–8960.

Chen, C. Y., Bertozzi, C., Zou, Z., Yuan, L., Lee, J. S., Lu, M., et al. (2012). Blood flow reprograms lymphatic vessels to blood vessels. *Journal of Clinical Investigation, 122*, 2006–2017.

Cheng, A. M., Rowley, B., Pao, W., Hayday, A., Bolen, J. B., & Pawson, T. (1995). Syk tyrosine kinase required for mouse viability and B-cell development. *Nature, 378*, 303–306.

Christou, C. M., Pearce, A. C., Watson, A. A., Mistry, A. R., Pollitt, A. Y., Fenton-May, A. E., et al. (2008). Renal cells activate the platelet receptor CLEC-2 through podoplanin. *Biochemical Journal, 411*, 133–140.

Colonna, M., Samaridis, J., & Angman, L. (2000). Molecular characterization of two novel C-type lectin-like receptors, one of which is selectively expressed in human dendritic cells. *European Journal of Immunology, 30*, 697–704.

Deppermann, C., Cherpokova, D., Nurden, P., Schulz, J. -N., Thielmann, I., Kraft, P., Vögtle, T., Kleinschnitz, C., Dütting, S., Krohne, G., Eming, S. A., Nurden, A. T., Eckes, B., Stoll, G., Stegner, D., & Bernhard Nieswandt, B. (2013). Gray platelet syndrome and defective thrombo-inflammation in Nbeal2-deficient mice. *Journal of Clinical Investigation, 123*, 3331–3342.

Finney, B. A., Schweighoffer, E., Navarro-Nunez, L., Benezech, C., Barone, F., Hughes, C. E., et al. (2012). CLEC-2 and Syk in the megakaryocytic/platelet lineage are essential for development. *Blood, 119*, 1747–1756.

Fu, J., Gerhardt, H., McDaniel, J. M., Xia, B., Liu, X., Ivanciu, L., et al. (2008). Endothelial cell O-glycan deficiency causes blood/lymphatic misconnections and consequent fatty liver disease in mice. *Journal of Clinical Investigation, 118*, 3725–3737.

Hagerling, R., Pollmann, C., Andreas, M., Schmidt, C., Nurmi, H., Adams, R. H., et al. (2013). A novel multistep mechanism for initial lymphangiogenesis in mouse embryos based on ultramicroscopy. *EMBO Journal, 32*, 629–644.

Hughes, C. E., Navarro-Nunez, L., Finney, B. A., Mourao-Sa, D., Pollitt, A. Y., & Watson, S. P. (2010). CLEC-2 is not required for platelet aggregation at arteriolar shear. *Journal of Thrombosis and Haemostasis, 8*, 2328–2332.

Ichise, H., Ichise, T., Ohtani, O., & Yoshida, N. (2009). Phospholipase Cgamma2 is necessary for separation of blood and lymphatic vasculature in mice. *Development, 136*, 191–195.

Kerrigan, A. M., Navarro-Nunez, L., Pyz, E., Finney, B. A., Willment, J. A., Watson, S. P., et al. (2012). Podoplanin-expressing inflammatory macrophages activate murine platelets via CLEC-2. *Journal of Thrombosis and Haemostasis, 10*, 484–486.

Kiefer, F., Brumell, J., Al-Alawi, N., Latour, S., Cheng, A., Veillette, A., et al. (1998). The Syk protein tyrosine kinase is essential for Fcγ receptor signaling in macrophages and neutrophils. *Molecular and Cellular Biology, 18*, 4209–4220.

Mahtab, E. A., Wijffels, M. C., Van Den Akker, N. M., Hahurij, N. D., Lie Venema, H., Wisse, L. J., et al. (2008). Cardiac malformations and myocardial abnormalities in podoplanin knockout mouse embryos: Correlation with abnormal epicardial development. *Developmental Dynamics, 237*, 847–857.

Martin-Villar, E., Megias, D., Castel, S., Yurrita, M. M., Vilaro, S., & Quintanilla, M. (2006). Podoplanin binds ERM proteins to activate RhoA and promote epithelial-mesenchymal transition. *Journal of Cell Science, 119*, 4541–4553.

Mourao-Sa, D., Robinson, M. J., Zelenay, S., Sancho, D., Chakravarty, P., Larsen, R., et al. (2011). CLEC-2 signaling via Syk in myeloid cells can regulate inflammatory responses. *European Journal of Immunology, 41*, 3040–3053.

Navarro, A., Perez, R. E., Rezaiekhaligh, M. H., Mabry, S. M., & Ekekezie, I. I. (2011). Polarized migration of lymphatic endothelial cells is critically dependent on podoplanin regulation of Cdc42. *American Journal of Physiology Lung Cellular and Molecular Physiology, 300*, L32–L42.

Osada, M., Inoue, O., Ding, G., Shirai, T., Ichise, H., Hirayama, K., et al. (2012). Platelet activation receptor CLEC-2 regulates blood/lymphatic vessel separation by inhibiting proliferation, migration, and tube formation of lymphatic endothelial cells. *Journal of Biological Chemistry, 287*, 22241–22252.

Peters, A., Pitcher, L. A., Sullivan, J. M., Mitsdoerffer, M., Acton, S. E., Franz, B., et al. (2011). Th17 cells induce ectopic lymphoid follicles in central nervous system tissue inflammation. *Immunity, 35*, 986–996.

Schacht, V., Ramirez, M. I., Hong, Y. K., Hirakawa, S., Feng, D., Harvey, N., et al. (2003). T1alpha/podoplanin deficiency disrupts normal lymphatic vasculature formation and causes lymphedema. *EMBO Journal, 22*, 3546–3556.

Sebzda, E., Hibbard, C., Sweeney, S., Abtahian, F., Bezman, N., Clemens, G., et al. (2006). Syk and Slp-76 mutant mice reveal a cell-autonomous hematopoietic cell contribution to vascular development. *Developmental Cell, 11*, 349–361.

Senis, Y. A., Tomlinson, M. G., Garcia, A., Dumon, S., Heath, V. L., Herbert, J., et al. (2007). A comprehensive proteomics and genomics analysis reveals novel transmembrane proteins in human platelets and mouse megakaryocytes including G6b-B, a novel immunoreceptor tyrosine-based inhibitory motif protein. *Molecular and Cellular Proteomics, 6*, 548–564.

Suzuki-Inoue, K., Fuller, G. L., Garcia, A., Eble, J. A., Pohlmann, S., Inoue, O., et al. (2006). A novel Syk-dependent mechanism of platelet activation by the C-type lectin receptor CLEC-2. *Blood, 107*, 542–549.

Suzuki-Inoue, K., Inoue, O., Ding, G., Nishimura, S., Hokamura, K., Eto, K., et al. (2010). Essential in vivo roles of the C-type lectin receptor CLEC-2: Embryonic/neonatal lethality of CLEC-2-deficient mice by blood/lymphatic misconnections and impaired thrombus formation of CLEC-2-deficient platelets. *Journal of Biological Chemistry, 285*, 24494–24507.

Suzuki-Inoue, K., Kato, Y., Inoue, O., Kaneko, M. K., Mishima, K., Yatomi, Y., et al. (2007). Involvement of the snake toxin receptor CLEC-2, in podoplanin-mediated platelet activation, by cancer cells. *Journal of Biological Chemistry, 282*, 25993–26001.

Tang, T., Li, L., Tang, J., Li, Y., Lin, W. Y., Martin, F., et al. (2010). A mouse knockout library for secreted and transmembrane proteins. *Nature Biotechnology, 28*, 749–755.

Turner, M., Mee, P. J., Costello, P. S., Williams, O., Price, A. A., Duddy, L. P., et al. (1995). Perinatal lethality and blocked B-cell development in mice lacking the tyrosine kinase Syk. *Nature, 378*, 298–302.

Uhrin, P., Zaujec, J., Breuss, J. M., Olcaydu, D., Chrenek, P., Stockinger, H., et al. (2010). Novel function for blood platelets and podoplanin in developmental separation of blood and lymphatic circulation. *Blood, 115*, 3997–4005.

Watson, S. P., Herbert, J. M., & Pollitt, A. Y. (2010). GPVI and CLEC-2 in hemostasis and vascular integrity. *Journal of Thrombosis and Haemostasis, 8*, 1456–1467.

Yang, Y., Garcia-Verdugo, J. M., Soriano-Navarro, M., Srinivasan, R. S., Scallan, J. P., Singh, M. K., et al. (2012). Lymphatic endothelial progenitors bud from the cardinal vein and intersomitic vessels in mammalian embryos. *Blood, 120*, 2340–2348.

Chapter 9
Interactions of Immune Cells and Lymphatic Vessels

Raghu P. Kataru, Yulia G. Lee, and Gou Young Koh

Abstract In addition to fluid and lipid absorption, immune cell trafficking has now become recognized as one of the major functions of the lymphatic system. Recently, several critical roles of the lymphatic vessels (LVs) in modulating immune reactions during both physiological and pathological conditions have been emerging. As LVs serve as conduits for immune cells, they come to closely interact with macrophages/monocytes, dendritic cells, and T and B lymphocytes. Accumulating evidences indicate that reciprocal interactions between the LVs and immune cells exist which cause considerable influence over the process of immune cell migration, LV growth, and ultimately certain immune reactions. This chapter discusses on the interactions of macrophages/monocytes and dendritic cells with peripheral LVs and on those of sinusoidal macrophages and T and B lymphocytes with lymph node LVs.

9.1 Introduction

In addition to their role in interstitial fluid absorption, lymphatic vessels (LVs) also function as conduits for immune cells migrating from peripheral tissues to draining lymph nodes (LNs) and subsequently to blood circulation. This route of immune cell migration via LVs requires close interaction between the two, which aids in initiation of host innate and adaptive immune responses against peripherally invading pathogens. Immune cells closely interact with LVs predominantly at two sites: peripheral tissues and draining LNs. In the peripheral tissues, antigen-bearing dendritic cells (DCs) and macrophages enter the LVs through discontinuous endothelial junctions. In the LNs, several types of myeloid and lymphoid cells along the

R.P. Kataru • Y.G. Lee • G.Y. Koh (✉)
Laboratory of Vascular Biology and Stem Cells, Graduate School of Biomedical Science and Engineering, Korea Advanced Institute of Science and Technology, 373-1, Guseong-dong, Daejeon 305-701, South Korea
e-mail: gykoh@kaist.ac.kr

F. Kiefer and S. Schulte-Merker (eds.), *Developmental Aspects of the Lymphatic Vascular System*, Advances in Anatomy, Embryology and Cell Biology 214,
DOI 10.1007/978-3-7091-1646-3_9,

LVs receive the incoming antigen-laden DCs, leading to initiation of immune responses. In addition, the LN sinusoidal macrophages, which adhere to the lymphatic endothelium, capture pathogens and prevent their systemic dissemination.

Interestingly, immune cells and LVs exhibit a reciprocal regulation of mutual migration and growth. Immune cells are thought to help in LV growth as a paracrine source of lymphangiogenic growth factors and by transdifferentiation into lymphatic endothelial cells (LECs), although the latter mechanism needs to be established with more extensive studies. In turn, LVs reciprocate by regulating immune cell migration through secretion of chemoattractant molecules. An understanding of the cross talk between immune cells and LVs during homeostasis and in disease states is essential for LV regulation in clinical settings. This chapter broadly summarizes the interaction of immune cells such as macrophages, DCs, and B and T lymphocytes with LVs and their reciprocal regulation, which results in a sustained immune reaction against pathogens.

9.2 Lymphatic Vessels: Macrophages and Lymphangiogenesis

Among the multiple types of immune cells that interact with LVs during development and in adulthood, macrophages are vital in LV regulation. Primarily, macrophage interaction with LVs occurs during tissue injury and inflammation. In response to inflammatory insults, bone marrow–derived myeloid cells (BMDMCs)/macrophages are recruited to the affected tissue, where they phagocytose the antigens, clear the debris, and migrate to the nearest draining LN via LVs, which facilitates inflammation resolution (Serhan and Savill 2005). This interaction between macrophages and LVs results in a reciprocal regulation of inflammation and lymphangiogenesis, in which LEC-derived chemotactic or adhesion molecules influence macrophage migration to LNs. In turn, macrophage-derived lymphangiogenic growth factors assist in LV remodeling via paracrine signaling pathways. Within heterogeneous macrophage populations in various tissue microenvironments, macrophages expressing the markers F4/80, CD11b, CD68, and lymphatic vessel endothelial hyaluronan receptor-1 (LYVE-1) are mainly identified as pro-lymphangiogenic cells because they secrete abundant lymphangiogenic growth factors, including VEGF-C, VEGF-D, and VEGF-A, in response to inflammatory stimuli (Ji 2012; Zumsteg and Christofori 2012).

9.2.1 Role of Macrophages in Developmental and Inflammatory Lymphangiogenesis

Lymphangiogenesis actively occurs during embryonic development, and LVs are formed around embryonic day (E) 9.5–10.5 in the mouse and embryonic week 6–7 in humans (Alitalo 2011). LECs expressing the homeobox transcription factor Prox-1 differentiate from a subset of venous endothelial cells of the cardinal vein and migrate in a VEGF-C/VEGFR-3-dependent manner to establish the primary lymph sac that later remodels into a delicate LV network throughout the body ((Alitalo et al. 2005; Oliver 2004) see also Chaps. 2 and 13 of this volume). Macrophages are implicated as a major source of the lymphangiogenic growth factors necessary for developmental lymphangiogenesis (Fig. 9.1a). The monocyte colony-stimulating factor encoding the *csf-1* gene is mutated in *op/op* mice, leading to loss of the macrophage population, and these *op/op* mice exhibit delayed postnatal LV development (Kubota et al. 2009). LYVE-1$^+$ macrophages are closely associated with mouse intestinal LVs during E16.5–20.5 (Kim et al. 2007), and monocytes expressing the cytoplasmic tyrosine kinase *Syk* can fuel lymphangiogenesis by secreting lymphangiogenic growth factors (Bohmer et al. 2010), indicating a strong developmental relationship between macrophages and lymphangiogenesis. In addition to a paracrine role in LV growth, macrophages are implicated in maintaining LV caliber in skin during development (Gordon et al. 2010).

Active lymphangiogenesis is also observed during inflammation (Kim et al. 2012). In this process, lymphangiogenesis is largely governed by lymphangiogenic growth factors secreted from macrophages in association with their corresponding receptors, VEGFR-3 and VEGFR-2 (Ji 2012; Kim et al. 2012; Tammela and Alitalo 2010; Zumsteg and Christofori 2012) (Fig. 9.1a). This association has been demonstrated in several experimental and preclinical models of inflammation including suture-induced corneal inflammation (Cursiefen et al. 2004), corneal transplantation (Maruyama et al. 2005), *Mycoplasma pulmonis*–induced chronic respiratory tract inflammation (Baluk et al. 2005), renal transplantation (Kerjaschki 2005), dermal wound healing (Maruyama et al. 2007), and lipopolysaccharide (LPS)-induced dermal and peritoneal inflammation (Kang et al. 2009; Kataru et al. 2009; Kim et al. 2009).

In addition to their paracrine roles in development and inflammation, macrophages can also induce neo-lymphangiogenesis via transdifferentiation into LECs and incorporation into LVs, based on the results of several studies (Fig. 9.1b). In the mouse embryo, CD45$^+$, F4/80$^+$, LYVE-1$^+$, and/or Prox-1$^+$ mesenchymal cells are in close contact with growing LVs (Buttler et al. 2006), hinting that macrophages can transdifferentiate into LECs or integrate with LVs during development. Macrophage transdifferentiation into LECs is widely reported during inflammation, including in-human renal transplants (Kerjaschki 2005), mouse corneal lymphatic vessels (Cursiefen et al. 2004; Maruyama et al. 2005), pancreatic islet inflammation and diabetes (Yin et al. 2011), and LPS-induced peritonitis (Hall et al. 2012).

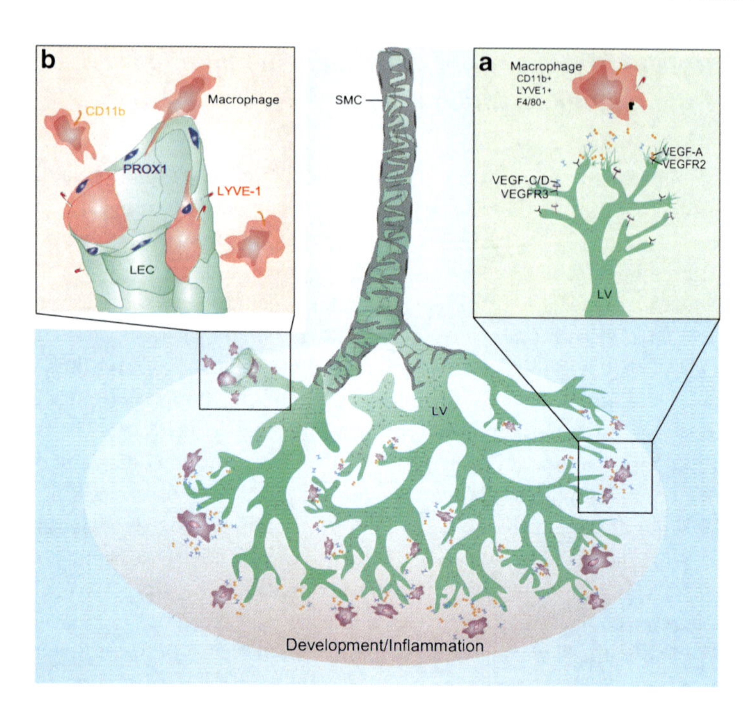

Fig. 9.1 Modes of macrophage-induced lymphangiogenesis. This schematic diagram illustrates two modes of macrophage-induced lymphangiogenesis. (**a**) Paracrine way: macrophages induce lymphangiogenesis primarily by secreting lymphangiogenic growth factors like VEGF-C/D and VEGF-A, which induce LEC sprouting, proliferation, and remodeling from the preexisting LVs in a VEGFR-3- or VEGFR-2-dependent manner. (**b**) Transdifferentiation or incorporation: macrophages also induce lymphangiogenesis by transdifferentiating into LECs or by being the source of LEC progenitor cells, which incorporate into growing LVs; however, this mechanism needs further investigation

Nevertheless, this novel concept of the transdifferentiation of macrophages into LECs requires further validation with precise and reliable assessments.

9.2.2 Sinusoidal Macrophages in LV

Lymphatic sinuses in LNs permanently tether a subset of macrophages known as LN-resident macrophages (Fig. 9.2a). These residential macrophages tightly adhere to LECs of lymphatic sinuses in both physiological and pathophysiological conditions (Gray and Cyster 2012). Based on location, LN-resident macrophages are classified into two subsets: subcapsular sinus macrophages (SSMs: $CD169^{hi}/CD11b^{hi}/CD11c^{lo}/F4/80^{-}$) and medullary sinus macrophages (MSMs: $CD169^{hi}/CD11b^{hi}/CD11c^{lo}/F4/80^{+}$) (Gray and Cyster 2012). SSMs are extraordinary in terms of their transcellular position between LECs, with the head portion residing in the lymphatic lumen, the long "tail" extended into the B cell follicle underneath,

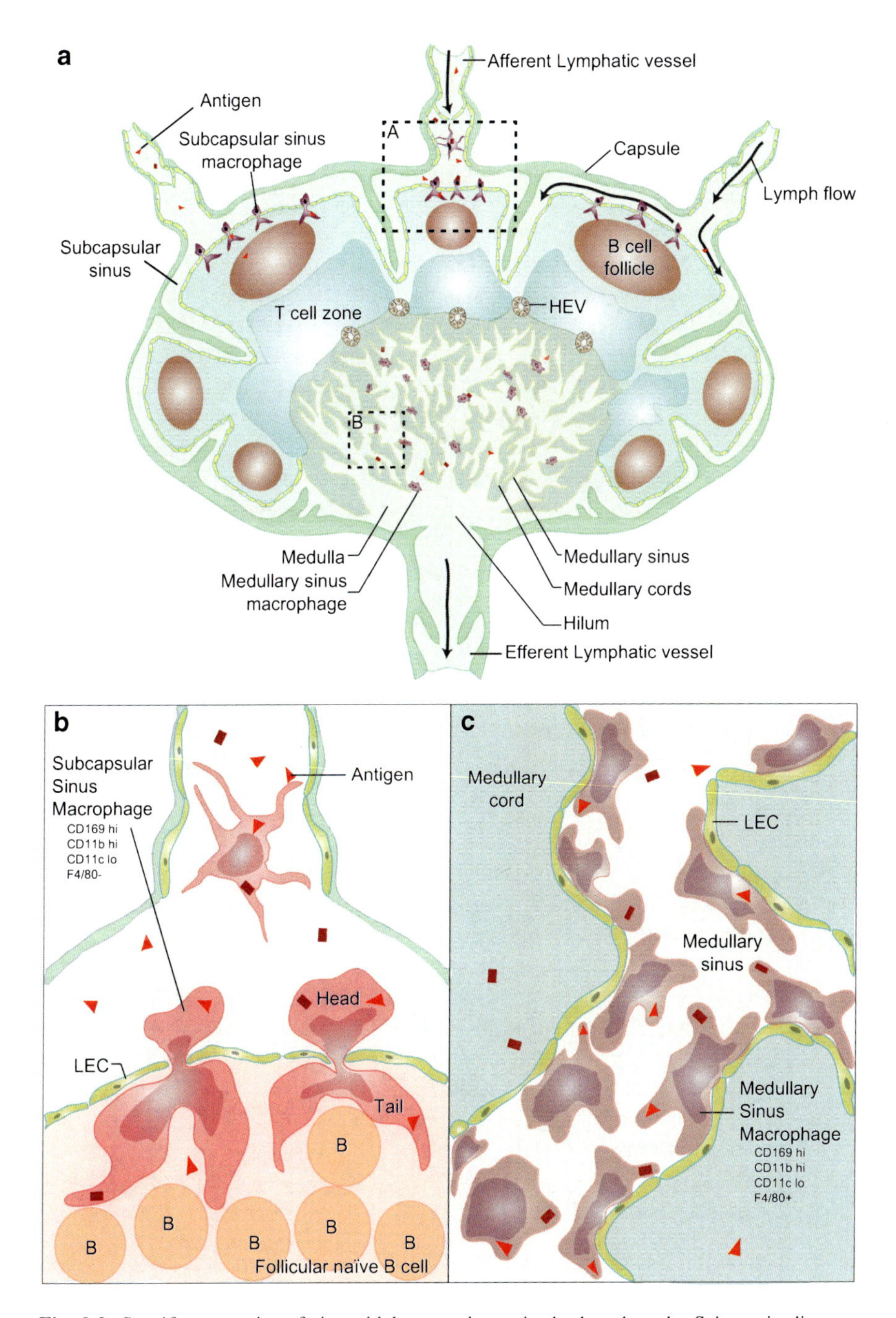

Fig. 9.2 Specific properties of sinusoidal macrophages in the lymph node. Schematic diagram illustrating closely the interaction of sinusoidal macrophages with lymphatic endothelium in the LN. (**a**) Overview of the LN structure with subcapsular and medullary lymphatic sinuses and their sinusoidal macrophages along with major T cell and B cell zones. *HEV* high endothelial venule. *Dotted boxes* are highlighted below. (**b**) A magnified view of the subcapsular sinus showing transcellularly positioned subcapsular sinus macrophages, which acquire antigens (*red triangles*

and the "neck" inserted between LECs (Fig. 9.2b). This transcellular positioning facilitates SSM acquisition of the lymph-borne antigen via the "head" and translocation and presentation of antigens to B cell follicles through the "tail" processes (Phan et al. 2009). Unlike SSMs, MSMs are adhered to LECs in the luminal side of medullary sinuses, carrying out phagocytosis and clearance of pathogens carried by lymph (Fig. 9.2c).

LVs can be hijacked by lymph-borne pathogens to gain systemic entry into a host unless they are obstructed in draining LNs (Kastenmuller et al. 2012). SSMs and MSMs stud the lymphatic sinus filter, capture and kill pathogens, and initiate humoral immunity. An abundance of research has been performed over the past decade on the involvement of SSMs and MSMs in antigen capture, prevention of trans-nodal dissemination of pathogens (Iannacone et al. 2010; Junt et al. 2007), and activation of B cell humoral immunity (Gonzalez et al. 2010; Phan et al. 2009). Despite increasing knowledge about the functional importance of LN macrophages, what remains unknown are the molecular cues with regard to their recruitment, binding, and interdependence with lymphatic sinuses in LN. To date, the pathogen barrier function of SSMs and MSMs has been elucidated mostly by their depletion using clodronate liposomes; however, it is important to determine whether or not disrupting the LEC and SSM/MSM interaction can destabilize the LN pathogen barrier function.

9.3 Mobilization of Dendritic Cells Through Lymph Vessels from Peripheral Organs to Draining Lymph Nodes

Migration of DCs via LVs, an upstream event of antigen presentation, is critical for adaptive immunity. Several chemokine ligand–receptor combinations play crucial roles as molecular traffic signals in DC migration through LVs (Alvarez et al. 2008; Randolph et al. 2005). A few DCs migrate to LNs in the resting state, whereas a large number of DCs migrate to LNs during inflammation. Augmented levels of the cytokines TNFα and IL-1β induced by local inflammatory stimulation are initial vital signals for expression of chemokine receptors and their corresponding ligands on DCs and LECs (Martin-Fontecha et al. 2003; Vigl et al. 2011). Among the many chemokine pathways for DC migration, CCR-7 and its ligand CCL21/CCL19 form the most prominent of these (Forster et al. 1999; Yanagihara et al. 1998). Indeed, CCL21 blockade results in compromised DC migration across LVs, as does CCR-7 deficiency in mice (Saeki et al. 1999). A recent report (Tal et al. 2011) demonstrated

Fig. 9.2 (continued) and *rectangles*) through their luminal "head" and present it to B cells through the abluminal "tail" to initiate humoral immunity. (**c**) A magnified view of the medullary sinus, showing adhered medullary sinus macrophages (MSMs) inside the sinus lumen. MSMs help capture and phagocytose incoming lymph-borne pathogens and prevent their trans-nodal dissemination

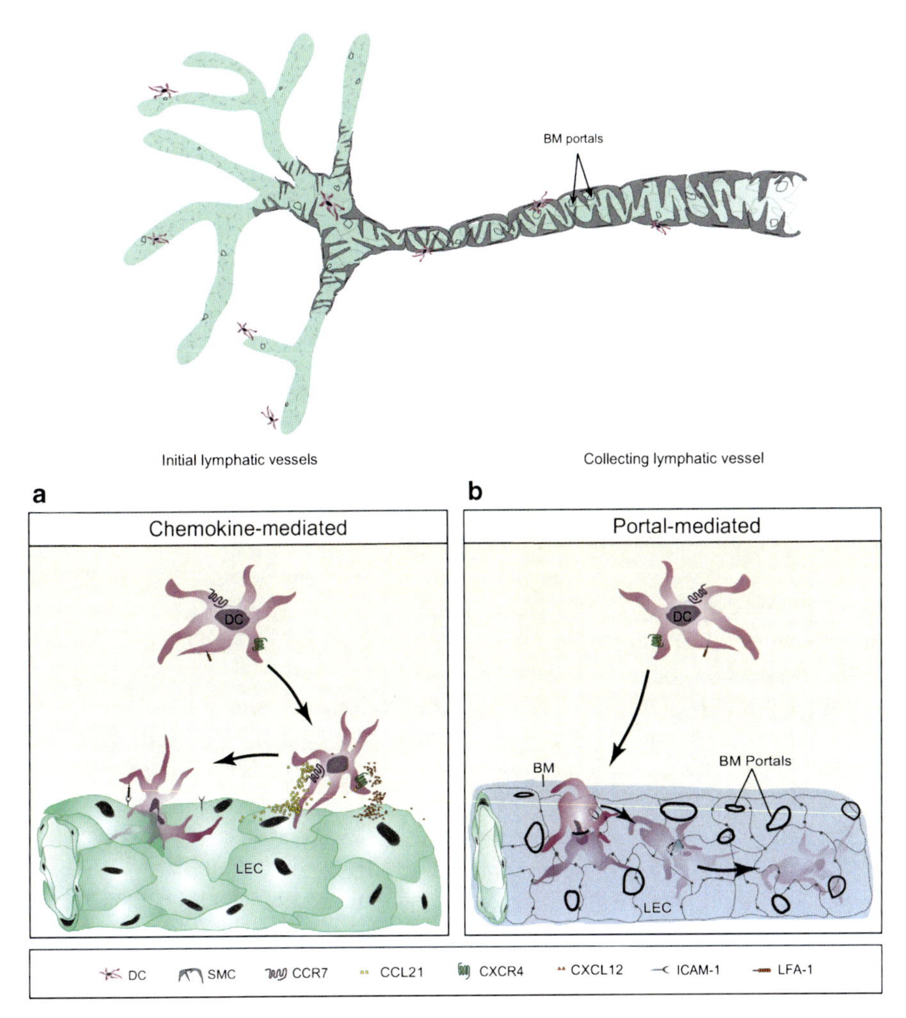

Fig. 9.3 Migration of DCs through LVs at peripheral organs. This schematic diagram illustrates alternative routes of DC chemoattraction and entry into peripheral LVs. (**a**) Chemokine-mediated: Activated DCs express chemokine receptors like CCR-7 and CXCR-4 and are strongly chemoattracted toward their respective ligand CCL21 and CXCL12 gradients around peripheral LVs. DCs transiently adhere to LECs through LFA-1/ICAM-1 interaction before their entry into the lymphatic lumen through discontinuous endothelial junctions. (**b**) Portal-mediated: Activated DCs enter preexisting dilated portals/gaps in the LV basement membrane before entering into the lymphatic lumen through the LEC flap valves or discontinuous endothelial junctions

that LEC-secreted CCL21 is deposited along the basement membrane of initial LVs, and as DCs migrate, they dock on the deposited CCL21 before transmigrating across the lymphatic endothelium. In addition to the CCL21/CCR-7 axis, DC migration is also enhanced through engagement of CXCR4-expressing DCs with CXCL12-expressing dermal LVs (Kabashima et al. 2007) (Fig. 9.3a). Furthermore, accumulating evidence indicates that VEGF-C/VEGFR-3 on the activated DCs and LECs is largely involved in DC migration across the LVs in LNs. Blocking VEGFR-3 signaling during cardiac allografts (Nykanen et al. 2010) and corneal transplants (Chen et al. 2004) leads to ablation of DC migration from the transplants to host LNs.

In the periphery, DCs enter LVs by transmigration across LEC junctions. DC entry via LEC junctions requires an interaction between DCs and several adhesion molecules on LECs. Prior to the LV entry, DC adherence is assisted by adhesion molecules such as ICAM-1, VCAM-1, and E-selectin on LECs, which are upregulated by the inflammatory cytokines (Fig. 9.3a). Previous studies (Johnson et al. 2006; Vigl et al. 2011) accordingly demonstrated that blocking ICAM-1 and VCAM-1 profoundly inhibits DC adhesion and transmigration across the lymphatic endothelium. A new mode of the lymphatic entry of DCs has been postulated, which involves preformed portals or gaps on the LV basement membrane. In this model, DCs directly enter portals of the initial LV basement membrane, which are mechanically dilated upon DC arrival, and penetrate the underlying LEC layer (Pflicke and Sixt 2009) (Fig. 9.3b) through discontinuous VE-cadherin junctions ((Baluk et al. 2007) and Chap. 4 of this issue). Considering the presence of discontinuous basement membrane on initial LVs, DC entry into LV lumen is a combination of, first, entering the basement membrane portals and, later, transmigration across LEC junctions.

After entry of DCs into the LV lumen, DCs migrate to draining LNs via afferent LVs. Although passive intra-lymphatic migration of DCs by lymphatic flow and transmural pressure has been proposed (Miteva et al. 2010), recent findings indicate involvement of active intra-lymphatic migration of DCs by Rho-associated protein kinase-regulated crawling (Nitschke et al. 2012) and lamellipodial extension-mediated crawling (Tal et al. 2011). After traversing through afferent LVs, DCs arrive at the subcapsular sinus of LNs and migrate down to the underlying paracortex region along a CCL21 gradient (Forster et al. 1999). Thus, starting from the initial entry, transit through LVs, and exit into LNs, DCs constantly are in close association with LVs. Controlling DC interaction with LVs at different sites could have different outcomes in various pathological conditions.

9.4 Role of B and T Cells in Lymphangiogenesis

B cells (or B lymphocytes) are a major group of lymphocytes that play a vital role in humoral immunity. Because of their large numbers in LNs, B cells usually orchestrate lymphangiogenesis within them. Angeli et al. (2006) found that B cells induce LN lymphangiogenesis during inflammation by secreting VEGF-A, resulting in increased intra-nodal DC mobilization and optimized immune activation. Moreover, B cell-specific over-expression of VEGF-A, even in the absence of inflammation, can induce LN lymphangiogenesis (Shrestha et al. 2010). These processes could be therapeutically exploited for quick adaptive immunity and efficient intranodal vaccine delivery.

T cells also comprise a major component of lymphocyte population in the paracortex region of LNs. Interestingly, LVs are easily found in the subcapsular and medullary regions but are scarce or absent in the paracortex region. This simple observation led to the identification of a negative regulatory role of T cells in LN LVs, which is mainly mediated by T cell-secreted interferon gamma (IFN-γ) (Kataru et al. 2011). IFN-γ exerts an anti-lymphatic effect by downregulating the major lymphatic players including Prox-1, resulting in decreased proliferation, migration, and tube formation (Kataru et al. 2011). These findings suggest a novel anti-lymphangiogenic role for T cells in balancing the pro-lymphatic drive of B cells and macrophages to maintain LV homeostasis in LNs.

A recent study using a mouse tail model of lymphedema (Zampell et al. 2012a) confirms the anti-lymphangiogenic role of T cells and IFN-γ in the peripheral tissue. Subsets of $CD4^+$ T cells (Zampell et al. 2012b) and Th2 cells secreting IL-4/IL-13 (Avraham et al. 2013) are anti-lymphangiogenic, causing tissue fibrosis and lymphatic stasis in the mouse tail model of lymphedema. Further understanding of the roles of different T cell subsets and their cytokines in lymphangiogenesis and lymphatic function would be helpful in the search for cures for secondary lymphedema caused by injuries and inflammation.

9.5 Implications and Future Directions

A growing body of evidence has established that immune cells closely interact with LVs during various pathophysiological processes, resulting in pro- and anti-lymphangiogenic effects and DC migration. Lymphangiogenesis and inflammatory disorders are highly interconnected, and regulation of lymphangiogenesis via modulation of macrophage-derived lymphangiogenic growth factors can be a potential strategy for treating chronic and acute inflammatory disorders, as several clinical and preclinical studies already have indicated. Moreover, B cell-induced LN lymphangiogenesis can be exploited clinically to promote DC migration or efficient vaccine delivery into LNs. As a corollary, dampening DC migration from the allograft site to host LNs by blocking the CCL21/CCR-7 or VEGF-C/VEGFR-3 pathways holds immense potential for preventing graft rejection after transplantation. Finally, depleting T cells or blocking T cell-secreted cytokines such as IFNγ, IL-4, and IL-13 seems to be a promising strategy for treating secondary lymphedema. Despite rapid progress in our knowledge related to immune cell and LV interactions using rodent models, translating this knowledge base into clinical therapies requires results from more clinically oriented studies that reflect precise human disease conditions.

Acknowledgments This work was supported by the grant (R2009-0079390 and 2011-0019268, GYK) of the National Research Foundation (NRF) funded by the MEST, Korea. We apologize to the many authors whose important work could not be cited because of space restrictions. The author has no conflicting financial interest.

References

Alitalo, K. (2011). The lymphatic vasculature in disease. *Nature Medicine, 17*, 1371–1380.

Alitalo, K., Tammela, T., & Petrova, T. V. (2005). Lymphangiogenesis in development and human disease. *Nature, 438*, 946–953.

Alvarez, D., Vollmann, E. H., & von Andrian, U. H. (2008). Mechanisms and consequences of dendritic cell migration. *Immunity, 29*, 325–342.

Angeli, V., Ginhoux, F., Llodra, J., Quemeneur, L., Frenette, P. S., Skobe, M., et al. (2006). B cell-driven lymphangiogenesis in inflamed lymph nodes enhances dendritic cell mobilization. *Immunity, 24*, 203–215.

Avraham, T., Zampell, J. C., Yan, A., Elhadad, S., Weitman, E. S., Rockson, S. G., et al. (2013). Th2 differentiation is necessary for soft tissue fibrosis and lymphatic dysfunction resulting from lymphedema. *FASEB Journal, 27*, 1114–1126.

Baluk, P., Fuxe, J., Hashizume, H., Romano, T., Lashnits, E., Butz, S., et al. (2007). Functionally specialized junctions between endothelial cells of lymphatic vessels. *Journal of Experimental Medicine, 204*, 2349–2362.

Baluk, P., Tammela, T., Ator, E., Lyubynska, N., Achen, M. G., Hicklin, D. J., et al. (2005). Pathogenesis of persistent lymphatic vessel hyperplasia in chronic airway inflammation. *Journal of Clinical Investigation, 115*, 247–257.

Bohmer, R., Neuhaus, B., Buhren, S., Zhang, D., Stehling, M., Bock, B., et al. (2010). Regulation of developmental lymphangiogenesis by Syk(+) leukocytes. *Developmental Cell, 18*, 437–449.

Buttler, K., Kreysing, A., von Kaisenberg, C. S., Schweigerer, L., Gale, N., Papoutsi, M., et al. (2006). Mesenchymal cells with leukocyte and lymphendothelial characteristics in murine embryos. *Developmental Dynamics, 235*, 1554–1562.

Chen, L., Hamrah, P., Cursiefen, C., Zhang, Q., Pytowski, B., Streilein, J. W., et al. (2004). Vascular endothelial growth factor receptor-3 mediates induction of corneal alloimmunity. *Nature Medicine, 10*, 813–815.

Cursiefen, C., Chen, L., Borges, L. P., Jackson, D., Cao, J., Radziejewski, C., et al. (2004). VEGF-A stimulates lymphangiogenesis and hemangiogenesis in inflammatory neovascularization via macrophage recruitment. *Journal of Clinical Investigation, 113*, 1040–1050.

Forster, R., Schubel, A., Breitfeld, D., Kremmer, E., Renner-Muller, I., Wolf, E., et al. (1999). CCR7 coordinates the primary immune response by establishing functional microenvironments in secondary lymphoid organs. *Cell, 99*, 23–33.

Gonzalez, S. F., Lukacs-Kornek, V., Kuligowski, M. P., Pitcher, L. A., Degn, S. E., Kim, Y. A., et al. (2010). Capture of influenza by medullary dendritic cells via SIGN-R1 is essential for humoral immunity in draining lymph nodes. *Nature Immunology, 11*, 427–434.

Gordon, E. J., Rao, S., Pollard, J. W., Nutt, S. L., Lang, R. A., & Harvey, N. L. (2010). Macrophages define dermal lymphatic vessel calibre during development by regulating lymphatic endothelial cell proliferation. *Development, 137*, 3899–3910.

Gray, E. E., & Cyster, J. G. (2012). Lymph node macrophages. *Journal of Innate Immunity, 4*, 424–436.

Hall, K. L., Volk-Draper, L. D., Flister, M. J., & Ran, S. (2012). New model of macrophage acquisition of the lymphatic endothelial phenotype. *PLoS One, 7*, e31794.

Iannacone, M., Moseman, E. A., Tonti, E., Bosurgi, L., Junt, T., Henrickson, S. E., et al. (2010). Subcapsular sinus macrophages prevent CNS invasion on peripheral infection with a neurotropic virus. *Nature, 465*, 1079–1083.

Ji, R. C. (2012). Macrophages are important mediators of either tumor- or inflammation-induced lymphangiogenesis. *Cellular and Molecular Life Sciences, 69*, 897–914.

Johnson, L. A., Clasper, S., Holt, A. P., Lalor, P. F., Baban, D., & Jackson, D. G. (2006). An inflammation-induced mechanism for leukocyte transmigration across lymphatic vessel endothelium. *Journal of Experimental Medicine, 203*, 2763–2777.

Junt, T., Moseman, E. A., Iannacone, M., Massberg, S., Lang, P. A., Boes, M., et al. (2007). Subcapsular sinus macrophages in lymph nodes clear lymph-borne viruses and present them to antiviral B cells. *Nature, 450*, 110–114.

Kabashima, K., Shiraishi, N., Sugita, K., Mori, T., Onoue, A., Kobayashi, M., et al. (2007). CXCL12-CXCR4 engagement is required for migration of cutaneous dendritic cells. *American Journal of Pathology, 171*, 1249–1257.

Kang, S., Lee, S. P., Kim, K. E., Kim, H. Z., Memet, S., & Koh, G. Y. (2009). Toll-like receptor 4 in lymphatic endothelial cells contributes to LPS-induced lymphangiogenesis by chemotactic recruitment of macrophages. *Blood, 113*, 2605–2613.

Kastenmuller, W., Torabi-Parizi, P., Subramanian, N., Lammermann, T., & Germain, R. N. (2012). A spatially-organized multicellular innate immune response in lymph nodes limits systemic pathogen spread. *Cell, 150*, 1235–1248.

Kataru, R. P., Jung, K., Jang, C., Yang, H., Schwendener, R. A., Baik, J. E., et al. (2009). Critical role of CD11b + macrophages and VEGF in inflammatory lymphangiogenesis, antigen clearance, and inflammation resolution. *Blood, 113*, 5650–5659.

Kataru, R. P., Kim, H., Jang, C., Choi, D. K., Koh, B. I., Kim, M., et al. (2011). T lymphocytes negatively regulate lymph node lymphatic vessel formation. *Immunity, 34*, 96–107.

Kerjaschki, D. (2005). The crucial role of macrophages in lymphangiogenesis. *Journal of Clinical Investigation, 115*, 2316–2319.

Kim, H., Kataru, R. P., & Koh, G. Y. (2012). Regulation and implications of inflammatory lymphangiogenesis. *Trends in Immunology, 33*, 350–356.

Kim, K. E., Koh, Y. J., Jeon, B. H., Jang, C., Han, J., Kataru, R. P., et al. (2009). Role of CD11b + macrophages in intraperitoneal lipopolysaccharide-induced aberrant lymphangiogenesis and lymphatic function in the diaphragm. *American Journal of Pathology, 175*, 1733–1745.

Kim, K. E., Sung, H. K., & Koh, G. Y. (2007). Lymphatic development in mouse small intestine. *Developmental Dynamics, 236*, 2020–2025.

Kubota, Y., Takubo, K., Shimizu, T., Ohno, H., Kishi, K., Shibuya, M., et al. (2009). M-CSF inhibition selectively targets pathological angiogenesis and lymphangiogenesis. *Journal of Experimental Medicine, 206*, 1089–1102.

Martin-Fontecha, A., Sebastiani, S., Hopken, U. E., Uguccioni, M., Lipp, M., Lanzavecchia, A., et al. (2003). Regulation of dendritic cell migration to the draining lymph node: Impact on T lymphocyte traffic and priming. *Journal of Experimental Medicine, 198*, 615–621.

Maruyama, K., Asai, J., Ii, M., Thorne, T., Losordo, D. W., & D'Amore, P. A. (2007). Decreased macrophage number and activation lead to reduced lymphatic vessel formation and contribute to impaired diabetic wound healing. *American Journal of Pathology, 170*, 1178–1191.

Maruyama, K., Ii, M., Cursiefen, C., Jackson, D. G., Keino, H., Tomita, M., et al. (2005). Inflammation-induced lymphangiogenesis in the cornea arises from CD11b-positive macrophages. *Journal of Clinical Investigation, 115*, 2363–2372.

Miteva, D. O., Rutkowski, J. M., Dixon, J. B., Kilarski, W., Shields, J. D., & Swartz, M. A. (2010). Transmural flow modulates cell and fluid transport functions of lymphatic endothelium. *Circulation Research, 106*, 920–931.

Nitschke, M., Aebischer, D., Abadier, M., Haener, S., Lucic, M., Vigl, B., et al. (2012). Differential requirement for ROCK in dendritic cell migration within lymphatic capillaries in steady-state and inflammation. *Blood, 120*, 2249–2258.

Nykanen, A. I., Sandelin, H., Krebs, R., Keranen, M. A., Tuuminen, R., Karpanen, T., et al. (2010). Targeting lymphatic vessel activation and CCL21 production by vascular endothelial growth factor receptor-3 inhibition has novel immunomodulatory and antiarteriosclerotic effects in cardiac allografts. *Circulation, 121*, 1413–1422.

Oliver, G. (2004). Lymphatic vasculature development. *Nature Reviews Immunology, 4*, 35–45.

Pflicke, H., & Sixt, M. (2009). Preformed portals facilitate dendritic cell entry into afferent lymphatic vessels. *Journal of Experimental Medicine, 206*, 2925–2935.

Phan, T. G., Green, J. A., Gray, E. E., Xu, Y., & Cyster, J. G. (2009). Immune complex relay by subcapsular sinus macrophages and noncognate B cells drives antibody affinity maturation. *Nature Immunology, 10*, 786–793.

Randolph, G. J., Angeli, V., & Swartz, M. A. (2005). Dendritic-cell trafficking to lymph nodes through lymphatic vessels. *Nature Reviews Immunology, 5*, 617–628.

Saeki, H., Moore, A. M., Brown, M. J., & Hwang, S. T. (1999). Cutting edge: secondary lymphoid-tissue chemokine (SLC) and CC chemokine receptor 7 (CCR7) participate in the emigration pathway of mature dendritic cells from the skin to regional lymph nodes. *Journal of Immunology, 162*, 2472–2475.

Serhan, C. N., & Savill, J. (2005). Resolution of inflammation: The beginning programs the end. *Nature Immunology, 6*, 1191–1197.

Shrestha, B., Hashiguchi, T., Ito, T., Miura, N., Takenouchi, K., Oyama, Y., et al. (2010). B cell-derived vascular endothelial growth factor A promotes lymphangiogenesis and high endothelial venule expansion in lymph nodes. *Journal of Immunology, 184*, 4819–4826.

Tal, O., Lim, H. Y., Gurevich, I., Milo, I., Shipony, Z., Ng, L. G., et al. (2011). DC mobilization from the skin requires docking to immobilized CCL21 on lymphatic endothelium and intralymphatic crawling. *Journal of Experimental Medicine, 208*, 2141–2153.

Tammela, T., & Alitalo, K. (2010). Lymphangiogenesis: Molecular mechanisms and future promise. *Cell, 140*, 460–476.

Vigl, B., Aebischer, D., Nitschke, M., Iolyeva, M., Rothlin, T., Antsiferova, O., et al. (2011). Tissue inflammation modulates gene expression of lymphatic endothelial cells and dendritic cell migration in a stimulus-dependent manner. *Blood, 118*, 205–215.

Yanagihara, S., Komura, E., Nagafune, J., Watarai, H., & Yamaguchi, Y. (1998). EBI1/CCR7 is a new member of dendritic cell chemokine receptor that is up-regulated upon maturation. *Journal of Immunology, 161*, 3096–3102.

Yin, N., Zhang, N., Lal, G., Xu, J., Yan, M., Ding, Y., et al. (2011). Lymphangiogenesis is required for pancreatic islet inflammation and diabetes. *PLoS One, 6*, e28023.

Zampell, J. C., Avraham, T., Yoder, N., Fort, N., Yan, A., Weitman, E. S., et al. (2012a). Lymphatic function is regulated by a coordinated expression of lymphangiogenic and anti-lymphangiogenic cytokines. *American Journal of Physiology Cellular Physiology, 302*, C392–C404.

Zampell, J. C., Yan, A., Elhadad, S., Avraham, T., Weitman, E., & Mehrara, B. J. (2012b). CD4(+) cells regulate fibrosis and lymphangiogenesis in response to lymphatic fluid stasis. *PLoS One, 7*, e49940.

Zumsteg, A., & Christofori, G. (2012). Myeloid cells and lymphangiogenesis. *Cold Spring Harbor Perspect Medicine, 2*, a006494.

Chapter 10
Lymphatic Vessels in the Development of Tissue and Organ Rejection

Deniz Hos and Claus Cursiefen

Abstract The lymphatic vascular system—amongst other tasks—is critically involved in the regulation of adaptive immune responses as it provides an important route for APC trafficking to secondary lymphatic organs. In this context, the cornea, which is the transparent and physiologically avascular "windscreen" of the eye, has served as an excellent in vivo model to study the role of the blood and lymphatic vasculature in mediating allogenic immune responses after transplantation. Especially the mouse model of high-risk corneal transplantation, where corneal avascularity is abolished by a severe inflammatory stimulus prior to keratoplasty, allows for comparison to other transplantations performed in primarily vascularized tissues and solid organs. Using this model, we recently demonstrated that especially lymphatic vessels, but not blood vessels, define the high-risk status of vascularized corneas and that anti(lymph)angiogenic treatment significantly promotes corneal allograft survival. Since evidence for lymphangiogenesis and its potential association with graft rejection is nowadays also present in solid organ transplantation, studies are currently addressing the potential benefits of anti(lymph)angiogenic treatment as a novel therapeutic concept also in solid organ grafting with promising initial results.

10.1 Introduction

The lymphatic vascular system is essential for fluid homeostasis and the resorption of dietary fats and is also involved in the regulation of inflammation, immune surveillance, and induction of adaptive immunity (Alitalo 2011; Alitalo et al. 2005;

All authors declare no financial disclosures.

D. Hos • C. Cursiefen (✉)
Department of Ophthalmology, University of Cologne, Kerpener Strasse 62, 50924 Cologne, Germany
e-mail: claus.cursiefen@uk-koeln.de

F. Kiefer and S. Schulte-Merker (eds.), *Developmental Aspects of the Lymphatic Vascular System*, Advances in Anatomy, Embryology and Cell Biology 214,
DOI 10.1007/978-3-7091-1646-3_10,

Potente et al. 2011; Tammela and Alitalo 2010). Generally, for the initiation of adaptive immune responses to foreign antigens, mature antigen-presenting cells (APCs) with high expression of major histocompatibility (MHC) class II from the periphery must interact with and prime naïve, resident T lymphocytes in lymph nodes. In this context, lymphatic vessels provide important paths for APC trafficking and for access of soluble antigens to the secondary lymphatic organs (Alitalo 2011; Dana 2006; Randolph et al. 2005).

Immunity to foreign antigens is essential to maintain the integrity of the organism. After organ or tissue transplantation, however, immune responses are undesired as the organism critically depends on graft function. Although the role of the lymphatic vasculature has extensively been studied in the context of adaptive immunity (Randolph et al. 2005), its involvement in *allogenic* immune responses in particular is so far only partly characterized. The majority of our knowledge regarding the role of lymphatic vessels in transplantation immunology is deduced from grafting experiments at the cornea (Cursiefen et al. 2003). Corneal transplantation is the most common and the most successful form of (tissue) transplantation (Streilein et al. 1999). This extraordinary success is partially due to a noticeable and important characteristic of the cornea: the healthy cornea belongs to the very few avascular and thereby also immune-privileged tissues of the body (Cursiefen 2007).

Corneal avascularity, also termed the corneal "angiogenic privilege," is essential for proper vision and is actively maintained by several antiangiogenic mechanisms (Albuquerque et al. 2009; Ambati et al. 2006; Bock et al. 2013; Cursiefen et al. 2006a; Singh et al. 2013). Nevertheless, severe inflammatory processes can overwrite this physiological corneal avascularity and result in pathological ingrowth of both blood and lymphatic vessels into the cornea, thereby immolating the functionality of the tissue for the safety of the organism (Cursiefen et al. 2003; Cursiefen et al. 2006b). Thus, the cornea has served as a very commonly used model system to study mechanisms of hem- and lymphangiogenesis, also because the accessibility and transparency of the cornea allows simple ex vivo and in vivo imaging.

Corneal transplantation immunology differs from other tissue and solid organ grafting. The avascular cornea is immunologically privileged, which results in excellent corneal graft survival rates without previous HLA matching if corneal transplantation is performed because of noninflammatory and nonvascular diseases like keratoconus or Fuchs' endothelial dystrophy ("low-risk" corneal transplantation). However, if the corneal angiogenic and immunologic privilege is lost due to severe inflammation and corneal grafts are placed into prevascularized, no more immune-privileged recipient beds, survival rates significantly decrease even under aggressive systemic immunosuppression ("high-risk" corneal transplantation) (Cursiefen et al. 2003; Dana and Streilein 1996; Niederkorn 2010; Sano et al. 1995).

It is possible to experimentally abolish the corneal angiogenic and immunologic privilege in mice to generate prevascularized and no more immune-privileged recipient beds. This permits the comparison to other transplantations performed in primarily vascularized tissues and solid organs (Bock et al. 2013; Dana and Streilein 1996; Sano et al. 1995). For this purpose, sutures are placed into the murine corneal stroma, which lead to a strong inflammatory response that overcomes the corneal (lymph)angiogenic and immunological privilege (Dana and Streilein 1996). Subsequently, the cornea becomes invaded by both blood and lymphatic vessels and immune cells. Using this high-risk model of murine corneal transplantation, we and other groups could recently provide new mechanistic insights into the involvement of not only blood but especially lymphatic vessels in transplantation immunology (Albuquerque et al. 2009; Bock et al. 2013; Dietrich et al. 2010).

In the first part of this chapter, we will provide an overview of corneal physiology, the mechanisms of corneal avascularity, and pathological corneal neovascularization and will focus on the role of blood versus lymphatic vessels in corneal transplantation immunology. In the second part, recent findings regarding lymphatic vessel function in solid organ transplantation will be recapitulated. Finally, we will discuss implications of these experimental results and their current and potential translation into clinical practice and will propose future directions of lymphangiogenesis research in the field of transplantation immunology.

10.2 Lymphatic Vessels in Corneal Allograft Rejection

10.2.1 Anatomy of the Physiologically Avascular and Alymphatic Cornea

The transparent cornea is the outermost part and the major refractive element of the eye. It primarily consists of five layers: (1) a stratified squamous nonkeratinized epithelium at the outside; (2) the subjacent Bowman's layer; (3) a middle collagen-rich stromal layer; and the inner layers of (4) Descemet's membrane and (5) corneal endothelium. The epithelial surface of the cornea and the superjacent tear film form an important barrier to the outside environment and guard the eye from microbial invasion, chemical and toxic damage, and foreign bodies. The stroma of the cornea comprises approximately 90 % of the cornea's total thickness in humans and 60–70 % in mice. Corneal stromal collagen fibrils display a unique and highly periodical distribution that minimizes light scattering and permits transparency. Constant dehydration of the stroma by the carbonic anhydrase activity of the endothelial cells results in tightly packed collagen lamellae and ensures stromal clarity. Severe inflammation or endothelial cell loss can lead to fluid storage (corneal edema), which diminishes stromal compactness and usually causes reduction of corneal transparency.

The cornea is one of the few avascular tissues of the body and contains no blood and lymphatic vessels (Cursiefen 2007). The evolutionary highly conserved ability of the cornea to actively inhibit the ingrowths of vessels to maintain its unique structure and function is also called the corneal "angiogenic privilege". Recently, several mechanisms that contribute to this angiogenic privilege could be identified (see Sect. 10.2.2) (Albuquerque et al. 2009; Ambati et al. 2006; Cursiefen et al. 2006a; Singh et al. 2013). Despite of its avascular nature, the cornea relies on blood components to remain healthy. Thus, delivery of nutrients and subsequent clearance of metabolites are carried out by the tear film from the corneal surface, by the aqueous humor from the anterior chamber, and at the limbus from the lateral margin. The limbus is the transition zone where the transparent and avascular cornea fades into the opaque and vascularized sclera and conjunctiva. Limbal arterioles originate from a ring of connected blood vessels posterior of the limbus, form small loops, and sprout towards the cornea, representing the pericorneal blood vessel arcades. The limbal lymphatic vasculature also consists of a ring-shaped network connected to the conjunctival lymphatic vessels. In contrast to the limbal blood vasculature, it does not form arcades but rather consists of one main circumferential lymphatic vessel with blind-ending extensions directed towards the central cornea (Cursiefen et al. 1998; Cursiefen et al. 2002b; Hos et al. 2008a). In addition, the main circumferential lymphatic vessel also may show gaps in its limbal circle (Hos et al. 2008a). Furthermore, limbal lymphatic vessels can form valves and have lymphatic vessel endothelial hyaluronan receptor 1 (LYVE-1) negative sections, which might serve as entrance zones for immune cells like APCs or antigen itself (Nakao et al. 2012; Steven et al. 2011; Truong et al. 2011).

The limbal blood and lymphatic vascular system is not static and shows remarkable plasticity. Changes can be observed during inflammation as well as aging (Cursiefen et al. 2006b; Hos et al. 2008a). Whereas minor inflammatory stimuli (to which the cornea is constantly exposed due to its anatomical position) do not induce a hem- or lymphangiogenic response, severe and eye-threatening inflammation of the cornea can lead to a breakdown of its angiogenic privilege and result in pathological ingrowths of both blood and lymphatic vessels (Sect. 10.2.3) (Bock et al. 2013; Cursiefen 2007). In contrast, the limbal vasculature physiologically regresses with age: the area covered by the main circumferential limbal lymphatic vessel decreases and shows considerably longer gaps (Hos et al. 2008a). Furthermore, the amount of centrally directed extensions from the main circumferential limbal lymphatic vessel is reduced, which correlates with the ability to respond to inflammatory and angiogenic stimuli (Hos et al. 2008a; Regenfuss et al. 2010). As a result, (lymph)angiogenic potency significantly decreases with age.

10.2.2 "Corneal Angiogenic Privilege": How Corneal Avascularity Is Actively Maintained

Corneal avascularity is not a passive condition but rather is actively maintained by multiple molecular mechanisms (Cursiefen 2007). It is essential for vision and is therefore evolutionarily highly conserved.

Several potent antiangiogenic molecules like endostatin, thrombospondin-1 (TSP-1), thrombospondin-2 (TSP-2), pigment epithelium-derived factor, and tissue inhibitor of metalloproteinase-3 are derived by the extracellular matrix component of the corneal epithelial basement membrane (Armstrong and Bornstein 2003; Cursiefen et al. 2004c; Lin et al. 2001). These molecules exhibit multiple inhibitory functions like direct blockade of endothelial cell migration and proliferation as well as indirect interference with growth factor mobilization and binding. Especially TSP-1 seems to be a key molecule critical for the maintenance of corneal lymphatic avascularity. In this context, we could recently demonstrate that TSP-1 reduces macrophage-derived expression of prolymphangiogenic vascular endothelial growth factor C (VEGF-C) via CD36 signaling and inhibits accumulation of macrophages in the cornea (Cursiefen et al. 2011). Accordingly, TSP-1 knockout mice show increased VEGF-C expression and higher corneal macrophage numbers, which results in spontaneous and isolated ingrowth of lymphatic vessels into the cornea (Cursiefen et al. 2011).

In addition, the corneal epithelium plays an essential role in maintaining corneal avascularity by the expression of several anti(lymph)angiogenic cytokine traps, which bind the key angiogenic growth factors VEGF-A, VEGF-C, and VEGF-D. For instance, the cornea expresses soluble VEGF receptor-1 (sVEGFR-1), soluble VEGFR-2 (sVEGFR-2), and soluble VEGFR-3 (sVEGFR-3), which act as decoy receptors and sequester (lymph)angiogenic VEGFs (Albuquerque et al. 2009; Ambati et al. 2006; Singh et al. 2013). Furthermore, these soluble receptors are able to form heterodimers with membrane bound VEGF receptors and inactivate them. Suppression of these endogenous VEGF traps (e.g., by genetic disruption, neutralizing antibodies or RNA interference) has been shown to abolish corneal avascularity. Moreover, we could previously demonstrate that the corneal epithelium ectopically expresses membrane bound VEGFR-3, which is also able to bind (lymph)angiogenic ligands and thereby maintain corneal avascularity (Cursiefen et al. 2006a).

As a further antiangiogenic mechanism, the cornea expresses the inhibitory PAS domain protein (IPAS), which negatively regulates hypoxia-induced upregulation of VEGF to actively maintain its angiogenic privilege even under hypoxic conditions (Makino et al. 2001).

However, the corneal angiogenic privilege described in this chapter can be overcome by severe inflammatory processes. Mechanisms leading to the breakdown of the angiogenic privilege are described in the following chapter.

10.2.3 Pathological Hem- and Lymphangiogenesis in Corneal Inflammation

The avascular cornea usually does not respond with hem- or lymphangiogenesis to minor inflammatory stimuli since an angiogenic reaction would interfere with corneal transparency (Cursiefen 2007). Nonetheless, several severe and eye-threatening inflammatory conditions can overwhelm the above described corneal antiangiogenic mechanisms and lead to the ingrowth of pathologic blood and lymphatic vessels into the cornea (Bock et al. 2013; Cursiefen 2007). Vascularization causes loss of corneal transparency not only by itself but also by secondary effects such as fluid and lipid deposition and the influx of inflammatory cells, resulting in corneal edema.

Inflammatory corneal hem- and lymphangiogenesis is mediated by a variety of resident corneal tissue cells and recruited inflammatory cells (Azar 2006; Cursiefen et al. 2004b; Maruyama et al. 2005). The corneal epithelium is able to secrete proangiogenic growth factors after corneal wounding, and it has also been shown that corneal fibroblasts upregulate proangiogenic growth factor expression as response to proinflammatory stimuli like TNF-α or IL-1β (Xi et al. 2011). There are also reports demonstrating low level VEGF-A expression in the corneal endothelium. However, it is so far unclear whether this expression is involved in inflammatory corneal hem- and lymphangiogenesis (Philipp et al. 2000).

In addition to resident tissue cells, inflammatory cells, in particular macrophages, play a decisive role in the process of inflammation-driven corneal hem- and lymphangiogenesis (Cursiefen et al. 2004b; Maruyama et al. 2005). After being recruited into the tissue, macrophages become activated and secrete not only proinflammatory cytokines such as TNF-α and IL-1β but also the major hem- and lymphangiogenic growth factors VEGF-A, VEGF-C, and VEGF-D, leading to blood and lymphatic endothelial cell activation, migration, and proliferation (Cursiefen et al. 2004b). Additionally, as macrophages themselves also express the respective receptors, more macrophages are recruited into the inflamed site and the immune and vascular response is further amplified (Cursiefen et al. 2004b). Moreover, macrophages can not only release vast amounts of (lymph)angiogenic growth factors but also express lymphatic markers and become integral components of newly formed lymphatic vessels in the cornea (Maruyama et al. 2005; Salven et al. 2003). In this context, Maruyama et al. demonstrated that corneal CD11b+ macrophages are able to integrate into preexisting lymphatic vessels and are also capable of generating vessel-like structures de novo (Maruyama et al. 2005). These primitive structures initially show no connection to preexisting limbal lymphatic vessels, but can subsequently fuse with each other or connect with preexisting lymphatic vessels. The essential role of macrophages in inflammatory corneal hem- and lymphangiogenesis is further evidenced by the fact that depletion of macrophages, e.g., by clodronate liposomes almost completely prevents the growth of both blood and lymphatic vessels into the cornea (Cursiefen et al. 2004b).

Newly formed pathological corneal blood and lymphatic vessels also show substantial plasticity. We have previously determined the time course of lymphatic vessel regression after a short but severe inflammatory stimulus and observed that lymphatic vessel regression started earlier and was more pronounced than blood vessel regression. 6 months after the inflammatory insult, partially perfused blood vessels were still present, whereas lymphatic vessels were no longer detectable (Cursiefen et al. 2006b). This knowledge is clinically essential to optimize the timing of corneal transplantation in patients with vascularized corneas due to severe inflammation, as the presence of corneal blood and lymphatic vessels is the most important risk factor for immune-mediated graft rejections after performed corneal transplantation.

10.2.4 Importance of Hem- Versus Lymphangiogenesis in Corneal Transplant Immunology

Pathological corneal hem- and lymphangiogenesis not only results in loss of corneal transparency and reduced visual acuity but also significantly increases the rate of immune-mediated graft rejection after subsequent corneal transplantation (Cursiefen et al. 2003; Dana and Streilein 1996; Sano et al. 1995). Whereas allogeneic corneal grafts placed into an avascular, so-called "low-risk" recipient bed exhibit high survival rates without HLA matching, graft survival significantly decreases when grafts are placed into prevascularized, "high-risk" recipient beds, even under systemic immunosuppression. Thus, immune reactions in high-risk keratoplasties are a major unsolved clinical problem.

Reasons for the exceptional good survival prognosis of corneal "low-risk" grafts are: first, the angiogenic privilege of the cornea; and, second, the immune privilege of the cornea. Several conditions are responsible for the corneal immune privilege, e.g., low numbers of MHC class II positive APCs especially in the corneal center, reduced expression of MHC class I in the cornea, expression of Fas ligand (CD95L) and programmed death ligand 1 (PD-L1), which both block immune responses by inhibiting T cell proliferation and inducing T cell apoptosis, and anterior chamber-associated immune deviation (ACAID), which causes systemic immune tolerance characterized by antigen-specific downregulation of delayed-type hypersensitivity (DTH) responses (Niederkorn 2010; Streilein et al. 1999). It has been shown that the corneal immune privilege depends on its angiogenic privilege, in parts, because various mediators of inflammation are also modulators of (lymph)angiogenesis. Therefore, prevascularized corneas with blood and lymphatic vessels show a combined loss of the corneal angiogenic and immune privilege (Cursiefen et al. 2003; Dana and Streilein 1996; Sano et al. 1995).

Preoperatively existing corneal neovascularization predisposes corneal grafts to accelerated induction of alloimmunity and subsequent rejection, because donor and recipient immune cells have immediate access to both the blood and lymphatic vasculatures (Fig. 10.1). It is likely that corneal lymphatic vessels facilitate easier exit of APCs and antigens from the graft to regional lymph nodes, where enhanced allosensitization occurs (induction of adaptive alloimmunity) and may also alter the activation state of APCs during their trafficking. In contrast, corneal blood vessels provide a route of subsequent entry for immune effector cells like CD4+ alloreactive T cells into the graft (execution of adaptive alloimmunity; Fig. 10.1) (Cursiefen et al. 2003). Even in the low-risk setting of corneal transplantation without preexisting corneal blood and lymphatic vessels, moderate (mainly iatrogenic) corneal hem- and lymphangiogenesis develops after grafting and can even

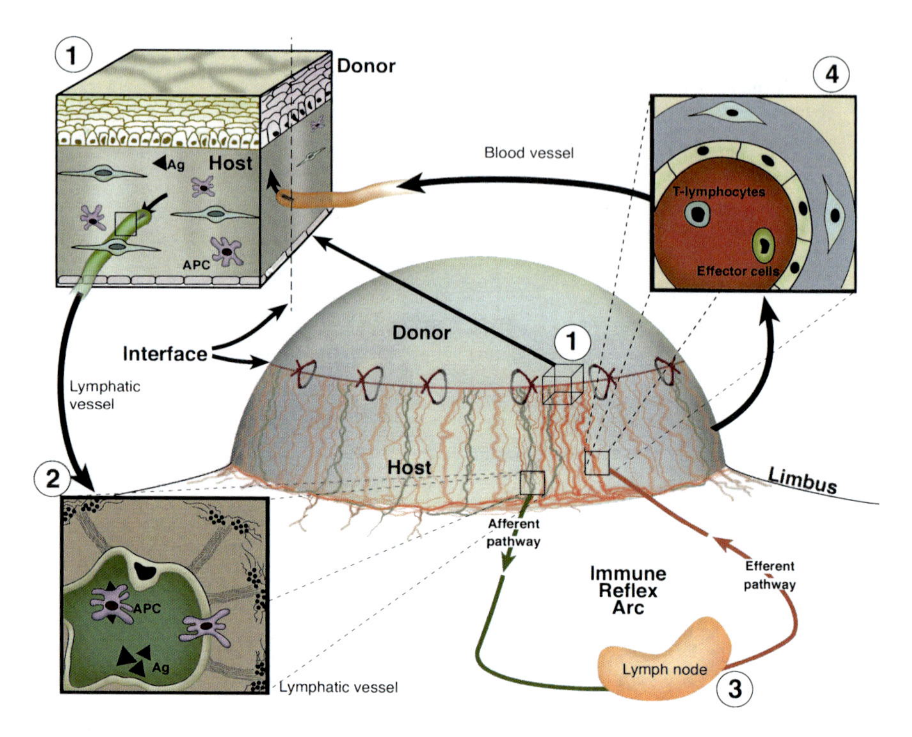

Fig. 10.1 Importance of blood and lymphatic vessels in high-risk corneal transplants for immune-mediated allograft rejection. (*1*) Magnification of the host-graft interface: blood (*red*) and lymphatic vessels (*green*) reach the host-graft interface. Antigen-presenting cells (APCs) and antigen (Ag) can leave the cornea via corneal lymphatic vessels (*2*) and reach the regional lymph nodes (*3*, afferent arm of the immune response); after stimulation of alloreactive T lymphocytes, these and other effector cells can reach the allograft via corneal blood vessels (*4*, efferent arm of the immune response). Modified from Cursiefen et al. (2003)

reach the host-graft interface under certain circumstances, also increasing the risk for subsequent immune-mediated graft rejection (Cursiefen et al. 2002a; Cursiefen et al. 2001; Dana et al. 1995).

Taken together, corneal neovascularization, the most important risk factor for immune reactions after transplantation, eases the connection between the graft and the regional lymph nodes (via lymphatic vessels), on the one hand, and alloreactive effector cells and the graft (via blood vessels), on the other hand (Fig. 10.1) (Cursiefen et al. 2003). Therefore, modulation of this "immune reflex pathway" seems a very reasonable approach to improve graft survival and to avoid graft rejection after both high- and low-risk corneal transplantations. Several studies have analyzed the potential effects of this interference on corneal graft survival so far. Notably, surgical excision of draining regional lymph nodes (cervical lymphadenectomy) in mice provided 100 % graft survival in low-risk recipient beds and 90 % graft survival in high-risk recipient beds (Yamagami and Dana 2001; Yamagami et al. 2002). Importantly, the improvement of graft survival by interfering with the local lymph node drainage seems to be due to the lack of allospecific immunity (immunologic "ignorance") and not due to tolerogenic mechanisms. Since then, we and other groups analyzed the feasibility of a "molecular lymphadenectomy," namely, a pharmacological blockade of the afferent and efferent arm of the immune reflex arc to improve corneal graft survival after transplantation. Decisively, combined inhibition of pathological corneal hem- and lymphangiogenesis effectively promoted graft survival, both in the low- and in the high-risk settings (see this chapter and Sect. 10.2.5) (Bachmann et al. 2008; Cursiefen et al. 2004a; Hos et al. 2008b).

Beyond that, the comparative importance of blood versus lymphatic vessels in corneal transplant rejection was not characterized until recently. Using different pharmacological inhibitors of lymphangiogenesis, namely, a small molecule antagonist of α5β1 integrin and a blocking antibody directed against VEGFR-3, we have previously shown that it is possible to solely block inflammatory corneal lymphangiogenesis without affecting inflammatory corneal hemangiogenesis (Bock et al. 2008b; Dietrich et al. 2007). By application of these inhibitors after corneal suture placement, we generated different recipient beds prior to corneal grafting: (1) avascular beds ("low-risk"); (2) hem- and lymphvascularized beds ("high-risk"); and (3) only hem- but not lymphvascularized beds after treatment with integrin α5β1 inhibitor or anti-VEGFR-3 ("alymphatic high-risk"; Fig. 10.2). Comparing graft survival rates after corneal transplantation, we observed that grafts placed into alymphatic but hemvascularized recipient beds had a similar graft survival compared to grafts placed into completely avascular, low-risk recipient beds, whereas the preexistence of lymphatic vessels in the high-risk setting significantly diminished corneal graft survival (Fig. 10.2). Therefore, we could demonstrate that lymphatic vessels and not blood vessels primarily define the high-risk status of prevascularized recipients (Dietrich et al. 2010). These results could recently be reproduced by administration of soluble VEGFR-2, which inhibits lymphatic vessel formation without influencing blood vessel growth. Treatment with soluble VEGFR-2 enhanced corneal allograft survival, although the recipients contained blood but not lymphatic vessels prior to grafting (Albuquerque et al. 2009).

In conclusion, these findings demonstrate that lymphatic vessels are a major risk factor for corneal graft rejection and seem to play a superior role in the mediation of alloimmune responses compared to blood vessels. These results may have important implications beyond corneal transplantation also in solid organ grafting, as solid organs depend much more on an intact blood vasculature to maintain organ function than the cornea.

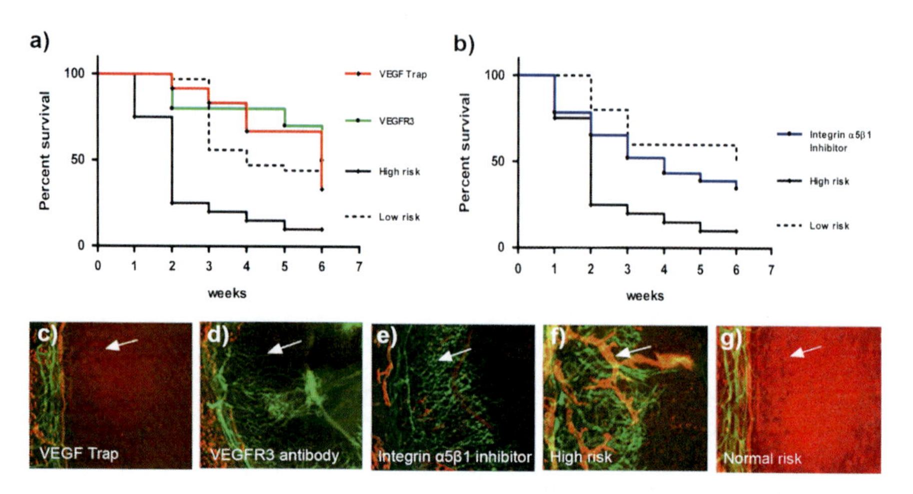

Fig. 10.2 Lymphatic vessels, but not blood vessels, primarily define the high-risk status of prevascularized corneal hosts. Corneal allograft survival was compared between hem- and lymphvascularized ("high-risk"; (**a**) and (**b**) [*black line*]) and avascular recipient controls ("low-risk"; (**a**) and (**b**) [*dotted line*]). Graft survival was significantly better when transplants were placed into recipient beds lacking lymphatic vessels ("alymphatic high-risk"; generated by the administration of VEGF-Trap, anti-VEGFR-3 or integrin α5β1 inhibitor) compared to recipient beds with lymphatic vessels present at the time of transplantation (VEGF-Trap vs. high-risk: $p < 0.0001$, *red line*; anti-VEGFR-3 vs. high-risk: $p < 0.0002$, *green line*; integrin α5β1 inhibitor vs. high-risk: $p < 0.032$, *blue line*; Kaplan-Meyer survival curve). Generation of different recipient beds prior to corneal grafting: pathologic corneal neovascularization in the recipient bed was induced by corneal suture placement and mice were treated with VEGF-Trap (**c**; resulting in no blood or lymphatic vessels but low inflammation present in the recipient bed at the time of transplantation), anti-VEGFR-3, or integrin α5β1 inhibitor (resulting in no lymphatic but only blood vessels present in the recipient bed at the time of transplantation; (**d**) and (**e**), respectively); (**f**) "high-risk" recipient (blood and lymphatic vessels present in the recipient bed at the time of transplantation); (**g**) "low-risk" recipient (no blood or lymphatic vessels in the recipient bed at the time of transplantation); *green*: blood vessels, CD31; *red*: lymphatic vessels, LYVE-1; *arrow*: prevascularized cornea. Modified from Dietrich et al. (2010)

10.2.5 Promotion of Corneal Graft Survival by Targeting Lymphangiogenesis as a Novel Therapeutic Concept

As already described in the previous chapters, it is generally accepted that corneal neovascularization is the main risk factor for immune-mediated graft rejections after corneal transplantation. The main mechanisms are: first, easier access of activated APCs and antigenic material from the graft to the regional lymph nodes via lymphatic vessels and possible modification of the APC activation state during entry into and passage through afferent lymphatics; second, sensitization and activation of alloreactive immune cells in the lymph nodes; and third, infiltration of alloreactive immune effector cells via blood vessels into the graft and subsequent rejection (Fig. 10.1). Therefore, targeting pathological corneal hem- and lymphangiogenesis to disrupt the afferent and efferent arm of the immune reflex arc was thought to be an effective therapeutic approach to reduce immune responses after corneal grafting (Cursiefen et al. 2003). For this reason, we and others investigated this concept of "pharmacological lymphadenectomy" in the murine model of allogenic corneal transplantation.

Initially, we analyzed the effect of an inhibition of hem- and lymphangiogenesis in the low-risk setting of corneal transplantation, where no blood and lymphatic vessels are present at the time of grafting, but have been shown to grow towards the graft and reach the graft-host interface within days after performed transplantation. Blockade of hem- and lymphangiogenesis was achieved by administration of VEGF-Trap, which selectively neutralizes VEGF-A and was previously shown to potently inhibit pathological corneal hem- and also lymphangiogenesis (Cursiefen et al. 2004b). Blockade of hem- and lymphangiogenesis directly after corneal transplantation resulted in a significant increase of corneal graft survival, showing for the first time that the novel therapeutic concept of immune modulation by anti (lymph)angiogenic therapy was successful (Cursiefen et al. 2004a).

However, as especially the high-risk situation of corneal transplantation is the major unsolved clinical problem, we analyzed the impact of a postoperative inhibition of hem- and lymphangiogenesis in corneas where pathological blood and lymphatic vessels were already present before corneal grafting. In this context, we could show that additional hem- and lymphangiogenesis occurred after transplantation and that postoperative inhibition of VEGF-A (by VEGF-Trap) efficiently diminished this additional neovascularization. Moreover, this blockade significantly increased corneal graft survival (Bachmann et al. 2008). Therefore, pharmacological inhibition of (lymph)angiogenesis even in the high-risk setting of corneal grafting is a novel and effective approach to improve graft survival in already prevascularized corneas, presenting a promising therapeutic tool to prevent immune-mediated graft rejection in these clinically challenging eyes.

In patients presenting with prevascularized corneas secondary to inflammation, the usually occurring partial regression of corneal neovascularization should be awaited before transplantation, if possible, and grafting should be performed during an interval of low or absent inflammation. To study the potential impact of corneal revascularization inhibition in previously inflamed and vascularized corneas on graft survival, we induced inflammatory corneal hem- and lymphangiogenesis in mice by corneal suture placement, awaited vessel regression for 8 months, and then performed allogenic corneal transplantations. Blockade of the reformation of partially regressed corneal blood and lymphatic vessels after grafting (by administration of VEGF-Trap) significantly promoted corneal graft survival, proving that the postoperative blockade of corneal hem- and lymphangiogenesis is beneficial also in this clinically highly relevant situation (Bachmann et al. 2009).

Taken together, anti(lymph)angiogenic therapy is a new therapeutic concept to improve graft survival after corneal transplantation, and translational approaches to adopt this novel treatment strategy into clinical practice have started. In this regard, we could previously show that bevacizumab, a recombinant, humanized, monoclonal antibody directed against VEGF-A and approved by the US Food and Drug Administration for the treatment of several metastatic carcinomas, inhibits not only corneal hem- but also lymphangiogenesis (Bock et al. 2007). Based on these findings, several case series and clinical trials have recently demonstrated that anti-VEGF treatment can reliably inhibit corneal vessel growth and also induce partial corneal vessel regression in patients (Bock et al. 2008a; Cheng et al. 2012; Ferrari et al. 2013). Furthermore, fine-needle vessel coagulation by diathermy in combination with topical use of bevacizumab also has been shown to effectively induce regression of mature corneal vessels prior to corneal grafting, presenting an effective method to "downgrade" prevascularized high-risk recipient beds to low-risk recipient beds (Koenig et al. 2012).

In addition, insulin receptor substrate-1 (IRS-1) has recently been recognized as an important intracellular signaling molecule involved in corneal neovascularization (Andrieu-Soler et al. 2005). Blockade of IRS-1 signaling by aganirsen eye drops, an antisense oligonucleotide against IRS-1, has been shown to effectively inhibit both corneal hem- and lymphangiogenesis (Hos et al. 2011). Aganirsen has since then successfully been tested as eye drops in patients and has entered clinical phase III trials (Cursiefen et al. 2009).

As already mentioned in the previous chapters, specific blockade of transplantation-associated lymphangiogenesis would be therapeutically beneficial especially in solid organ transplantation, as this approach would keep the blood vasculature intact and maintain organ perfusion. In addition to selective inhibition of lymphatic vessel progression, current research effort investigates the feasibility of a selective *regression* of lymphatic vessels prior to grafting.

10.3 Lymphatic Vessels in Solid Organ Rejection

In this chapter, we will first give a brief overview about general mechanisms of APC trafficking to lymph nodes and different ways of antigen presentation to T lymphocytes. Second, we will summarize recent findings in the field of lymphangiogenesis and solid organ transplantation. Finally, we will provide examples for anti(lymph)angiogenic treatment strategies in various organ systems.

10.3.1 Antigen Presenting Cell Trafficking and Antigen Delivery to Lymph Nodes Through the Lymphatic Vascular System

Alloantigen delivery via APCs to secondary lymphoid organs and their encounter with T lymphocytes is an obligatory precondition for the initiation of an adaptive alloimmune response after transplantation. Although some studies could show that small soluble antigens do not necessarily require APCs to reach the lymph nodes (Roozendaal et al. 2008), it seems that a highly effective T lymphocyte response still depends on the transfer of antigen-loaded APCs to secondary lymphoid organs (Alvarez et al. 2008). The importance of the secondary lymphoid organs in transplant immunology is reflected by the fact that surgical lymphadenectomy significantly promotes corneal allograft survival (Yamagami and Dana 2001; Yamagami et al. 2002). This is further confirmed by the observation that alymphoplastic (*aly/aly*) mice lacking secondary lymphoid organs do not reject allogenic skin and organ grafts (Lakkis et al. 2000). The transfer of antigen-loaded APCs, in particular dendritic cells (DCs), to lymph nodes mainly occurs via afferent lymphatic vessels. Therefore, to reach the regional lymph nodes, DCs must migrate towards the lymphatic vessel, cross the lymphatic endothelium, and subsequently enter the lymphatic vessel lumen. Although the exact molecular mechanisms of this multistep process remain so far only partially characterized, several factors have already been shown to facilitate DC trafficking to lymph nodes.

In this context, Johnson and colleagues demonstrated that DC migration depends on the expression of intercellular adhesion molecule-1 (ICAM-1) and vascular cell adhesion molecule-1 (VCAM-1) by lymphatic endothelial cells (Johnson et al. 2006). Lymphatic endothelial cells upregulate both ICAM-1 and VCAM-1 expression in response to the proinflammatory cytokines TNF-α and IL-1, thus significantly enhancing DC transfer to lymph nodes under inflammatory conditions. Furthermore, junctional adhesion molecule 1 (JAM1), another cell adhesion molecule, which is expressed by both DCs and lymphatic endothelial cells, regulates DC migration to the regional lymph nodes. It has been shown that DC transfer to lymph nodes is increased in mice deficient in JAM1, suggesting an inhibitory function of JAM1 on DC trafficking (Cera et al. 2004). In addition, studies demonstrated that the cell adhesion molecules common lymphatic endothelial and vascular receptor-1 (CLEVER-1) and mannose receptor 1 also regulate APC transfer to the lymph nodes (Salmi and Jalkanen 2005).

Attraction to chemokines is another major component of appropriate DC trafficking to lymph nodes. In this regard, the CC chemokine receptor 7 (CCR7) – CC chemokine ligand 19 (CCL19)/CC chemokine ligand 21 (CCL21) system seems to be the key pathway for lymph node homing of DCs (Forster et al. 2008; Forster et al. 1999). Mature DCs, which upregulate MHC class II and further costimulatory molecules necessary for efficient T cell priming, also express CCR7, which results in enhanced DC motility. CCR7 binds its ligands CCL19 and CCL21, which are expressed by stromal cells and lymphatic endothelial cells (Alvarez et al. 2008; Forster et al. 2008; Luther et al. 2000). By secreting CCL21, lymphatic endothelial cells generate a chemotactic gradient that successively guides DCs to the draining lymph nodes (Johnson and Jackson 2010). The importance of the CCR7 – CCL19/CCL21 axis is evidenced by the observation that CCR7-knockout mice or mice with CCR7-deficient DCs display substantially decreased DC migration to lymph nodes (Forster et al. 1999; Martin-Fontecha et al. 2003; Ohl et al. 2004).

Although some of the underlying cellular and molecular mechanisms of APC trafficking to secondary lymphatic organs have recently been deciphered, this complex process is still not fully understood and many questions remain unanswered. Nevertheless, with recent technological advances in intravital microscopy and multiphoton imaging, it is nowadays possible to directly visualize DC migration, DC-lymphatic interaction, and DC trafficking to lymph nodes in vivo (Lindquist et al. 2004; Martinez-Corral et al. 2012; Steven et al. 2011), which will surely shed light on further mechanisms in the near future.

When activated APCs encounter with T lymphocytes in the secondary lymphatic organs, the differentiation and proliferation of T cells is induced, and the adaptive immune response to nonself-antigen is triggered (Dana 2006; Karpanen and Alitalo 2008; Randolph et al. 2005). How APCs allosensitize T lymphocytes will be the focus of the following chapter.

10.3.2 Direct Versus Indirect Allorecognition

One of the main functions of the immune system is to distinguish between self and nonself and to destroy nonself as it presents a menace for the organism. In the context of allogenic tissue and organ transplantation, this implies that the graft is recognized as nonself and a subsequent immune response is generated, which potentially leads to graft rejection. As already mentioned, it is necessary for the generation of the alloimmune response that activated APCs encounter with T lymphocytes in lymph nodes and present donor-derived antigen in combination with costimulatory molecules, which leads to the activation of the T cells.

In the transplant setting, two distinct mechanisms for the presentation of alloantigen by APCs to recipient T cells exist, namely, the direct and indirect pathways of alloantigen recognition (Afzali et al. 2008; Gould and Auchincloss 1999). These pathways are mainly determined by which APCs present the alloantigen to the recipient T cells in secondary lymphoid organs: in the direct pathway, intact donor MHC molecules on the surface of donor passenger APCs are recognized by recipient T cells, whereas in the indirect pathway, donor antigen (mainly donor MHC molecules but also minor antigen) which is processed by recipient APCs and presented as peptides on self-MHC molecules is recognized by recipient T cells (Afzali et al. 2008; Gould and Auchincloss 1999).

It is so far not entirely clear, which allorecognition pathway is the major contributor to alloimmune responses, and it seems that this also depends on the respective tissue and organ system. Generally, direct allorecognition seems to be more dominant at early time points directly after grafting and may account for acute rejections, whereas indirect allorecognition seems to become more important at later time points and may account for chronic rejections (Afzali et al. 2008; Liu et al. 1996; Pietra et al. 2000). This is at least partially explained by the fact that donor APCs are gradually reduced and disappear with time.

10.3.3 Association of Lymphangiogenesis and Solid Organ Rejection: Evidence from Kidney and Cardiac Transplantation

Although the outcome of solid organ transplantation has markedly improved over the past decades, mainly because of the availability and use of immunosuppressive drugs, the long-term fate of transplanted organs still remains at risk of chronic rejection. Furthermore, most of the currently used immunosuppressive drugs inhibit the immune system unspecifically and display numerous side effects, demanding drugs with more specific interference and selective blockade of alloimmunity. Assuming that APC trafficking to the secondary lymphoid organs is an early and essential step in the induction of alloimmune responses also in solid organ grafting, the idea emerged that therapeutic intervention with this afferent arm of the immune reflex arc could be a rational way of modulating alloimmunity. However, for a long time, there was no definitive evidence for the involvement of lymphatic vessels in solid organ rejection, partly because reliable and specific markers for lymphatic endothelial cells were missing. Initial evidence that lymphangiogenesis occurs in solid organ grafting emerged in the fields of renal and cardiac transplantation (Kerjaschki et al. 2006; Kerjaschki et al. 2004; Nykänen et al. 2010).

Former studies had demonstrated that an inflammatory infiltrate is present in chronically rejected human kidney and heart transplants. In this context, Kerjaschki and coworkers were able to show that renal grafts contained numerous highly proliferating lymphatic vessels which were associated with organized cellular infiltrates containing CCR7+ cells and VEGF-C producing macrophages (Kerjaschki et al. 2006; Kerjaschki et al. 2004). Moreover, lymphatic vessels in the transplants expressed CCL21, the main chemokine for the guidance of CCR7+ APCs to the draining lymph nodes besides CCL19. Therefore, this study provided first direct evidence for lymphangiogenesis in renal transplants and further suggested dendritic cell chemoattraction by newly formed lymphatic vessels in the transplanted kidneys (Kerjaschki et al. 2006; Kerjaschki et al. 2004). Furthermore, the same group showed that in addition to preexisting endothelial cells also circulating, recipient-derived lymphatic endothelial progenitor cells, probably macrophages, contributed to inflammation-associated lymphangiogenesis in the grafts (Kerjaschki et al. 2006). These were the first studies proving that newly formed lymphatic vessels were present in solid organ transplants.

By using rat and mouse heart transplantation models, Nykänen and colleagues investigated the potential involvement of lymphangiogenesis in cardiac grafting (Nykänen et al. 2010). This study demonstrated that chronic rejection-associated inflammation induced lymphatic vessel growth also in cardiac allografts and that graft-infiltrating macrophages and CD4+ T cells were the main sources of lymphangiogenic VEGF-C and VEGF-D expression. Interestingly, in contrast to studies from Kerjaschki et al. in renal grafts, most of the lymphatic endothelial cells in cardiac allografts were donor-derived but not recipient-derived endothelial progenitor cells. Furthermore, this study also demonstrated that newly formed lymphatic vessels produce CCL21, indicating that transplantation-associated lymphangiogenesis in cardiac transplants is functional in attracting activated CCR7+ APCs to secondary lymphoid tissue and inducing an alloimmune response (Nykänen et al. 2010).

10.3.4 Promotion of Allograft Survival by Targeting Lymphangiogenesis in Various Organ Systems

Since the occurrence of lymphangiogenesis in rejected renal and cardiac transplants was proven, additional studies in other organ systems also confirmed the presence of newly formed lymphatic vessels and their association with chronic allograft rejection, e.g., in allogenic liver and lung transplantation (Dashkevich et al. 2010; Ishii et al. 2010). Based on these findings, the impact of anti(lymph)angiogenic treatment strategies on allograft survival was evaluated. In this chapter, we will present recent studies, which already demonstrated beneficial effects of this novel anti(lymph)angiogenic concept in various organ systems.

10.3.4.1 Renal Transplantation

The presence of increased lymphangiogenesis in chronically rejected renal allografts and evidence for its involvement in CCR7/CCL21-mediated APC transfer to secondary lymphoid organs implicates an important role of lymphatic vessels in the mediation of allograft responses (Kerjaschki et al. 2006; Kerjaschki et al. 2004). Therefore, Palin et al. analyzed the impact of sirolimus, an inhibitor of mammalian target of rapamycin (mTOR) that was previously shown to potently block lymphangiogenesis, on renal allograft function (Palin et al. 2013). In a rat model of allogenic renal transplantation, sirolimus inhibited VEGF-C/VEGFR-3-mediated lymphatic vessel formation in renal allografts and significantly reduced the amount of inflammatory infiltrates, intragraft fibrosis, and the rate of graft rejection in treated animals when compared to cyclosporine A (Palin et al. 2013). The authors concluded that the inhibition of lymphangiogenesis is a possible mechanism contributing to the protective effects of sirolimus on renal allograft function. However, as sirolimus is also an effective immunosuppressive drug inhibiting IL-2 and IL-4 induced T cell proliferation as well, studies interfering more specifically with renal grafting-associated lymphangiogenesis are still missing.

10.3.4.2 Cardiac Transplantation

The already mentioned study by Nykänen et al. (see Sect. 10.3.3), which demonstrated the occurrence of CCL21-producing lymphatic vessels in cardiac allografts, also investigated whether VEGFR-3 inhibition influences cardiac allograft function (Nykänen et al. 2010). Therefore, rat cardiac allograft recipients were treated with an adenoviral vector encoding soluble VEGFR-3-Ig (Ad.VEGFR-3-Ig), which traps VEGF-C and VEGF-D. Although the quantity of lymphangiogenesis in the allograft was not significantly affected, treatment with Ad.VEGFR-3-Ig significantly prolonged cardiac allograft survival and inhibited lymphatic vessel activation. Ad. VEGFR-3-Ig decreased allograft CCL21 expression and diminished the number of graft-infiltrating macrophages and CD8+ T cells, whereas the number of DCs within the grafts remained unaltered. Furthermore, DC recruitment to the spleen was markedly reduced by VEGFR-3 inhibition. The authors therefore concluded that blockade of VEGFR-3 affected CCL21-mediated DC migration to lymphatic vessels and their homing to the secondary lymphoid organs, similar to the observed effects of VEGFR-3 inhibition in the model of corneal transplantation (Chen et al. 2004), suggesting VEGFR-3 as a potential therapeutic target to reduce immune-mediated allograft responses and to improve graft survival after cardiac transplantation (Nykänen et al. 2010).

10.3.4.3 Islet Cell Transplantation

Recently, Källskog et al. could demonstrate that in islet grafts, substantial lymphangiogenesis develops after islet cell transplantation, although endogenous islets are usually devoid of lymphatic vessels (Källskog et al. 2006). This observation was further analyzed by Yin and coworkers, who showed that treatment with sunitinib, an anti(lymph)angiogenic receptor tyrosine kinase inhibitor, or specific blockade of VEGFR-3 signaling by a VEGFR-3-neutralizing antibody decreased intragraft and lymph node lymphangiogenesis and significantly delayed or even prevented islet cell allograft rejection (Yin et al. 2011).

10.4 Conclusions and Future Directions

Lymphangiogenesis research was neglected for a long time because reliable and specific markers for lymphatic endothelium were missing. However, after more specific markers for lymphatic endothelial cells were discovered in the last decades, the cellular and molecular mechanisms underlying lymphatic vessel growth are now being rapidly deciphered. Furthermore, recent advances in imaging techniques such as intravital microscopy and multiphoton imaging nowadays allow to almost directly visualize DC trafficking and lymphatic vessels and their interaction in vivo (Steven et al. 2011). In the context of lymphangiogenesis research, the cornea has served as an excellent in vivo model to study the lymphatic vascular system in transplant immunology. Especially the mouse model of high-risk corneal transplantation, where pathological blood and lymphatic vessels are present prior to grafting, permits comparison to transplantations performed in primarily vascularized tissues and organs. Several studies have demonstrated that lymphangiogenesis has essential functions in the induction of allogenic immune responses and is required for the trafficking of APCs and antigen to the secondary lymphoid organs, where allosensitization against the graft occurs. Using the mouse model of high-risk keratoplasty, we and other groups demonstrated that anti(lymph) angiogenic treatment significantly promotes corneal allograft survival (Bachmann et al. 2008; Bachmann et al. 2009; Dietrich et al. 2010). Furthermore, we were able to show that mainly lymphatic vessels, but not blood vessels, define the high-risk status of prevascularized recipients (Dietrich et al. 2010). Since evidence for lymphangiogenesis and its potential association with graft rejection is nowadays also present in solid organ grafts, studies are currently addressing the potential benefits of anti(lymph)angiogenic treatment as a novel therapeutic concept also in solid organ grafting with promising initial results. The fact that corneal allograft rejection depends mainly on lymph- but not hemangiogenesis (Dietrich et al. 2010) is of particular interest especially in solid organ grafting, as organ graft function critically depends on an unaltered blood vasculature. Therefore, recent studies are focusing on the feasibility of a specific blockade of lymphatic but not blood vessel growth. Additionally, selective (lymph)angioregressive treatment strategies (Koenig et al. 2012) would also be a further therapeutic approach to reduce the risk of graft rejection in tissue as well as solid organ grafting.

Acknowledgements The authors thank Birgit Regenfuss and Felix Bock (Department of Ophthalmology, University of Cologne) for helpful discussions and proofreading of the article. This work was supported by the German Research Foundation, Sonderforschungsbereich SFB 643 (B10), DFG Cu 47/4-1, and DFG Cu 47/6-1, and by the GEROK-Programme, University of Cologne (to DH).

References

Afzali, B., Lombardi, G., & Lechler, R. I. (2008). Pathways of major histocompatibility complex allorecognition. *Current Opinion in Organ Transplantation, 13*, 438–444.

Albuquerque, R. J., Hayashi, T., Cho, W. G., Kleinman, M. E., Dridi, S., Takeda, A., et al. (2009). Alternatively spliced vascular endothelial growth factor receptor-2 is an essential endogenous inhibitor of lymphatic vessel growth. *Nature Medicine, 15*, 1023–1030.

Alitalo, K. (2011). The lymphatic vasculature in disease. *Nature Medicine, 17*, 1371–1380.

Alitalo, K., Tammela, T., & Petrova, T. V. (2005). Lymphangiogenesis in development and human disease. *Nature, 438*, 946–953.

Alvarez, D., Vollmann, E. H., & von Andrian, U. H. (2008). Mechanisms and consequences of dendritic cell migration. *Immunity, 29*, 325–342.

Ambati, B. K., Nozaki, M., Singh, N., Takeda, A., Jani, P. D., Suthar, T., et al. (2006). Corneal avascularity is due to soluble VEGF receptor-1. *Nature, 443*, 993–997.

Andrieu-Soler, C., Berdugo, M., Doat, M., Courtois, Y., BenEzra, D., & Behar-Cohen, F. (2005). Downregulation of IRS-1 expression causes inhibition of corneal angiogenesis. *Investigative Ophthalmology & Visual Science, 46*, 4072–4078.

Armstrong, L. C., & Bornstein, P. (2003). Thrombospondins 1 and 2 function as inhibitors of angiogenesis. *Matrix Biology, 22*, 63–71.

Azar, D. T. (2006). Corneal angiogenic privilege: angiogenic and antiangiogenic factors in corneal avascularity, vasculogenesis, and wound healing (an American Ophthalmological Society thesis). *Transactions of the American Ophthalmological Society, 104*, 264–302.

Bachmann, B. O., Bock, F., Wiegand, S. J., Maruyama, K., Dana, M. R., Kruse, F. E., et al. (2008). Promotion of graft survival by vascular endothelial growth factor a neutralization after high-risk corneal transplantation. *Archives of Ophthalmology, 126*, 71–77.

Bachmann, B. O., Luetjen-Drecoll, E., Bock, F., Wiegand, S. J., Hos, D., Dana, R., et al. (2009). Transient postoperative vascular endothelial growth factor (VEGF)-neutralisation improves graft survival in corneas with partly regressed inflammatory neovascularisation. *The British Journal of Ophthalmology, 93*, 1075–1080.

Bock, F., Konig, Y., Kruse, F., Baier, M., & Cursiefen, C. (2008a). Bevacizumab (Avastin) eye drops inhibit corneal neovascularization. *Graefe's Archive for Clinical and Experimental Ophthalmology, 246*, 281–284.

Bock, F., Maruyama, K., Regenfuss, B., Hos, D., Steven, P., Heindl, L. M., et al. (2013). Novel anti (lymph)angiogenic treatment strategies for corneal and ocular surface diseases. *Progress in Retinal and Eye Research, 34*, 89–124.

Bock, F., Onderka, J., Dietrich, T., Bachmann, B., Kruse, F. E., Paschke, M., et al. (2007). Bevacizumab as a potent inhibitor of inflammatory corneal angiogenesis and lymphangiogenesis. *Investigative Ophthalmology & Visual Science, 48*, 2545–2552.

Bock, F., Onderka, J., Dietrich, T., Bachmann, B., Pytowski, B., & Cursiefen, C. (2008b). Blockade of VEGFR3-signalling specifically inhibits lymphangiogenesis in inflammatory corneal neovascularisation. *Graefe's Archive for Clinical and Experimental Ophthalmology, 246*, 115–119.

Cera, M. R., Del Prete, A., Vecchi, A., Corada, M., Martin-Padura, I., Motoike, T., et al. (2004). Increased DC trafficking to lymph nodes and contact hypersensitivity in junctional adhesion molecule-A-deficient mice. *The Journal of Clinical Investigation, 114*, 729–738.

Chen, L., Hamrah, P., Cursiefen, C., Zhang, Q., Pytowski, B., Streilein, J. W., et al. (2004). Vascular endothelial growth factor receptor-3 mediates induction of corneal alloimmunity. *Nature Medicine, 10*, 813–815.

Cheng, S. F., Dastjerdi, M. H., Ferrari, G., Okanobo, A., Bower, K. S., Ryan, D. S., et al. (2012). Short-term topical bevacizumab in the treatment of stable corneal neovascularization. *American Journal of Ophthalmology, 154*(940–948), e941.

Cursiefen, C. (2007). Immune privilege and angiogenic privilege of the cornea. *Chemical Immunology and Allergy, 92*, 50–57.

Cursiefen, C., Bock, F., Horn, F. K., Kruse, F. E., Seitz, B., Borderie, V., et al. (2009). GS-101 antisense oligonucleotide eye drops inhibit corneal neovascularization: Interim results of a randomized phase II trial. *Ophthalmology, 116*, 1630–1637.

Cursiefen, C., Cao, J., Chen, L., Liu, Y., Maruyama, K., Jackson, D., et al. (2004a). Inhibition of hemangiogenesis and lymphangiogenesis after normal-risk corneal transplantation by neutralizing VEGF promotes graft survival. *Investigative Ophthalmology & Visual Science, 45*, 2666–2673.

Cursiefen, C., Chen, L., Borges, L. P., Jackson, D., Cao, J., Radziejewski, C., et al. (2004b). VEGF-A stimulates lymphangiogenesis and hemangiogenesis in inflammatory neovascularization via macrophage recruitment. *The Journal of Clinical Investigation, 113*, 1040–1050.

Cursiefen, C., Chen, L., Dana, M. R., & Streilein, J. W. (2003). Corneal lymphangiogenesis: Evidence, mechanisms, and implications for corneal transplant immunology. *Cornea, 22*, 273–281.

Cursiefen, C., Chen, L., Saint-Geniez, M., Hamrah, P., Jin, Y., Rashid, S., et al. (2006a). Nonvascular VEGF receptor 3 expression by corneal epithelium maintains avascularity and vision. *Proceedings of the National Academy of Sciences of the United States of America, 103*, 11405–11410.

Cursiefen, C., Kuchle, M., & Naumann, G. O. (1998). Angiogenesis in corneal diseases: Histopathologic evaluation of 254 human corneal buttons with neovascularization. *Cornea, 17*, 611–613.

Cursiefen, C., Martus, P., Nguyen, N. X., Langenbucher, A., Seitz, B., & Kuchle, M. (2002a). Corneal neovascularization after nonmechanical versus mechanical corneal trephination for non-high-risk keratoplasty. *Cornea, 21*, 648–652.

Cursiefen, C., Maruyama, K., Bock, F., Saban, D., Sadrai, Z., Lawler, J., et al. (2011). Thrombospondin 1 inhibits inflammatory lymphangiogenesis by CD36 ligation on monocytes. *The Journal of Experimental Medicine, 208*, 1083–1092.

Cursiefen, C., Maruyama, K., Jackson, D. G., Streilein, J. W., & Kruse, F. E. (2006b). Time course of angiogenesis and lymphangiogenesis after brief corneal inflammation. *Cornea, 25*, 443–447.

Cursiefen, C., Masli, S., Ng, T. F., Dana, M. R., Bornstein, P., Lawler, J., et al. (2004c). Roles of thrombospondin-1 and -2 in regulating corneal and iris angiogenesis. *Investigative Ophthalmology & Visual Science, 45*, 1117–1124.

Cursiefen, C., Schlotzer-Schrehardt, U., Kuchle, M., Sorokin, L., Breiteneder-Geleff, S., Alitalo, K., et al. (2002b). Lymphatic vessels in vascularized human corneas: Immunohistochemical investigation using LYVE-1 and podoplanin. *Investigative Ophthalmology & Visual Science, 43*, 2127–2135.

Cursiefen, C., Wenkel, H., Martus, P., Langenbucher, A., Nguyen, N. X., Seitz, B., et al. (2001). Impact of short-term versus long-term topical steroids on corneal neovascularization after non-high-risk keratoplasty. *Graefe's Archive for Clinical and Experimental Ophthalmology, 239*, 514–521.

Dana, M. R. (2006). Angiogenesis and lymphangiogenesis-implications for corneal immunity. *Seminars in Ophthalmology, 21*, 19–22.

Dana, M. R., Schaumberg, D. A., Kowal, V. O., Goren, M. B., Rapuano, C. J., Laibson, P. R., et al. (1995). Corneal neovascularization after penetrating keratoplasty. *Cornea, 14*, 604–609.

Dana, M. R., & Streilein, J. W. (1996). Loss and restoration of immune privilege in eyes with corneal neovascularization. *Investigative Ophthalmology & Visual Science, 37*, 2485–2494.

Dashkevich, A., Heilmann, C., Kayser, G., Germann, M., Beyersdorf, F., Passlick, B., et al. (2010). Lymph angiogenesis after lung transplantation and relation to acute organ rejection in humans. *The Annals of Thoracic Surgery, 90*, 406–411.

Dietrich, T., Bock, F., Yuen, D., Hos, D., Bachmann, B. O., Zahn, G., et al. (2010). Cutting edge: Lymphatic vessels, not blood vessels, primarily mediate immune rejections after transplantation. *Journal of Immunology, 184*, 535–539.

Dietrich, T., Onderka, J., Bock, F., Kruse, F. E., Vossmeyer, D., Stragies, R., et al. (2007). Inhibition of inflammatory lymphangiogenesis by integrin alpha5 blockade. *The American Journal of Pathology, 171*, 361–372.

Ferrari, G., Dastjerdi, M. H., Okanobo, A., Cheng, S. F., Amparo, F., Nallasamy, N., et al. (2013). Topical ranibizumab as a treatment of corneal neovascularization. *Cornea, 32*(7), 992–997.

Forster, R., Davalos-Misslitz, A. C., & Rot, A. (2008). CCR7 and its ligands: Balancing immunity and tolerance. *Nature Reviews. Immunology, 8*, 362–371.

Forster, R., Schubel, A., Breitfeld, D., Kremmer, E., Renner-Muller, I., Wolf, E., et al. (1999). CCR7 coordinates the primary immune response by establishing functional microenvironments in secondary lymphoid organs. *Cell, 99*, 23–33.

Gould, D. S., & Auchincloss, H., Jr. (1999). Direct and indirect recognition: The role of MHC antigens in graft rejection. *Immunology Today, 20*, 77–82.

Hos, D., Bachmann, B., Bock, F., Onderka, J., & Cursiefen, C. (2008a). Age-related changes in murine limbal lymphatic vessels and corneal lymphangiogenesis. *Experimental Eye Research, 87*, 427–432.

Hos, D., Bock, F., Dietrich, T., Onderka, J., Kruse, F. E., Thierauch, K. H., et al. (2008b). Inflammatory corneal (lymph)angiogenesis is blocked by VEGFR-tyrosine kinase inhibitor

ZK 261991, resulting in improved graft survival after corneal transplantation. *Investigative Ophthalmology & Visual Science, 49*, 1836–1842.

Hos, D., Regenfuss, B., Bock, F., & Cursiefen, C. (2011). Blockade of insulin receptor substrate-1 inhibits corneal lymphangiogenesis. *Investigative Ophthalmology & Visual Science, 52*, 5778–5785.

Ishii, E., Shimizu, A., Kuwahara, N., Arai, T., Kataoka, M., Wakamatsu, K., et al. (2010). Lymphangiogenesis associated with acute cellular rejection in rat liver transplantation. *Transplantation Proceedings, 42*, 4282–4285.

Johnson, L. A., Clasper, S., Holt, A. P., Lalor, P. F., Baban, D., & Jackson, D. G. (2006). An inflammation-induced mechanism for leukocyte transmigration across lymphatic vessel endothelium. *The Journal of Experimental Medicine, 203*, 2763–2777.

Johnson, L. A., & Jackson, D. G. (2010). Inflammation-induced secretion of CCL21 in lymphatic endothelium is a key regulator of integrin-mediated dendritic cell transmigration. *International Immunology, 22*, 839–849.

Källskog, O., Kampf, C., Andersson, A., Carlsson, P. O., Hansell, P., Johansson, M., et al. (2006). Lymphatic vessels in pancreatic islets implanted under the renal capsule of rats. *American Journal of Transplantation, 6*, 680–686.

Karpanen, T., & Alitalo, K. (2008). Molecular biology and pathology of lymphangiogenesis. *Annual Review of Pathology, 3*, 367–397.

Kerjaschki, D., Huttary, N., Raab, I., Regele, H., Bojarski-Nagy, K., Bartel, G., et al. (2006). Lymphatic endothelial progenitor cells contribute to de novo lymphangiogenesis in human renal transplants. *Nature Medicine, 12*, 230–234.

Kerjaschki, D., Regele, H. M., Moosberger, I., Nagy-Bojarski, K., Watschinger, B., Soleiman, A., et al. (2004). Lymphatic neoangiogenesis in human kidney transplants is associated with immunologically active lymphocytic infiltrates. *Journal of the American Society of Nephrology, 15*, 603–612.

Koenig, Y., Bock, F., Kruse, F. E., Stock, K., & Cursiefen, C. (2012). Angioregressive pretreatment of mature corneal blood vessels before keratoplasty: Fine-needle vessel coagulation combined with anti-VEGFs. *Cornea, 31*, 887–892.

Lakkis, F. G., Arakelov, A., Konieczny, B. T., & Inoue, Y. (2000). Immunologic "ignorance" of vascularized organ transplants in the absence of secondary lymphoid tissue. *Nature Medicine, 6*, 686–688.

Lin, H. C., Chang, J. H., Jain, S., Gabison, E. E., Kure, T., Kato, T., et al. (2001). Matrilysin cleavage of corneal collagen type XVIII NC1 domain and generation of a 28-kDa fragment. *Investigative Ophthalmology & Visual Science, 42*, 2517–2524.

Lindquist, R. L., Shakhar, G., Dudziak, D., Wardemann, H., Eisenreich, T., Dustin, M. L., et al. (2004). Visualizing dendritic cell networks in vivo. *Nature Immunology, 5*, 1243–1250.

Liu, Z., Colovai, A. I., Tugulea, S., Reed, E. F., Fisher, P. E., Mancini, D., et al. (1996). Indirect recognition of donor HLA-DR peptides in organ allograft rejection. *The Journal of Clinical Investigation, 98*, 1150–1157.

Luther, S. A., Tang, H. L., Hyman, P. L., Farr, A. G., & Cyster, J. G. (2000). Coexpression of the chemokines ELC and SLC by T zone stromal cells and deletion of the ELC gene in the plt/plt mouse. *Proceedings of the National Academy of Sciences of the United States of America, 97*, 12694–12699.

Makino, Y., Cao, R., Svensson, K., Bertilsson, G., Asman, M., Tanaka, H., et al. (2001). Inhibitory PAS domain protein is a negative regulator of hypoxia-inducible gene expression. *Nature, 414*, 550–554.

Martinez-Corral, I., Olmeda, D., Dieguez-Hurtado, R., Tammela, T., Alitalo, K., & Ortega, S. (2012). In vivo imaging of lymphatic vessels in development, wound healing, inflammation, and tumor metastasis. *Proceedings of the National Academy of Sciences of the United States of America, 109*, 6223–6228.

Martin-Fontecha, A., Sebastiani, S., Hopken, U. E., Uguccioni, M., Lipp, M., Lanzavecchia, A., et al. (2003). Regulation of dendritic cell migration to the draining lymph node: Impact on T lymphocyte traffic and priming. *The Journal of Experimental Medicine, 198*, 615–621.

Maruyama, K., Ii, M., Cursiefen, C., Jackson, D. G., Keino, H., Tomita, M., et al. (2005). Inflammation-induced lymphangiogenesis in the cornea arises from CD11b-positive macrophages. *The Journal of Clinical Investigation, 115*, 2363–2372.

Nakao, S., Zandi, S., Faez, S., Kohno, R., & Hafezi-Moghadam, A. (2012). Discontinuous LYVE-1 expression in corneal limbal lymphatics: Dual function as microvalves and immunological hot spots. *FASEB Journal, 26*, 808–817.

Niederkorn, J. Y. (2010). High-risk corneal allografts and why they lose their immune privilege. *Current Opinion in Allergy and Clinical Immunology, 10*, 493–497.

Nykänen, A. I., Sandelin, H., Krebs, R., Keranen, M. A., Tuuminen, R., Karpanen, T., et al. (2010). Targeting lymphatic vessel activation and CCL21 production by vascular endothelial growth factor receptor-3 inhibition has novel immunomodulatory and antiarteriosclerotic effects in cardiac allografts. *Circulation, 121*, 1413–1422.

Ohl, L., Mohaupt, M., Czeloth, N., Hintzen, G., Kiafard, Z., Zwirner, J., et al. (2004). CCR7 governs skin dendritic cell migration under inflammatory and steady-state conditions. *Immunity, 21*, 279–288.

Palin, N. K., Savikko, J., & Koskinen, P. K. (2013). Sirolimus inhibits lymphangiogenesis in rat renal allografts, a novel mechanism to prevent chronic kidney allograft injury. *Transplant International, 26*, 195–205.

Philipp, W., Speicher, L., & Humpel, C. (2000). Expression of vascular endothelial growth factor and its receptors in inflamed and vascularized human corneas. *Investigative Ophthalmology & Visual Science, 41*, 2514–2522.

Pietra, B. A., Wiseman, A., Bolwerk, A., Rizeq, M., & Gill, R. G. (2000). CD4 T cell-mediated cardiac allograft rejection requires donor but not host MHC class II. *The Journal of Clinical Investigation, 106*, 1003–1010.

Potente, M., Gerhardt, H., & Carmeliet, P. (2011). Basic and therapeutic aspects of angiogenesis. *Cell, 146*, 873–887.

Randolph, G. J., Angeli, V., & Swartz, M. A. (2005). Dendritic-cell trafficking to lymph nodes through lymphatic vessels. *Nature Reviews. Immunology, 5*, 617–628.

Regenfuss, B., Onderka, J., Bock, F., Hos, D., Maruyama, K., & Cursiefen, C. (2010). Genetic heterogeneity of lymphangiogenesis in different mouse strains. *The American Journal of Pathology, 177*, 501–510.

Roozendaal, R., Mebius, R. E., & Kraal, G. (2008). The conduit system of the lymph node. *International Immunology, 20*, 1483–1487.

Salmi, M., & Jalkanen, S. (2005). Cell-surface enzymes in control of leukocyte trafficking. *Nature Reviews. Immunology, 5*, 760–771.

Salven, P., Mustjoki, S., Alitalo, R., Alitalo, K., & Rafii, S. (2003). VEGFR-3 and CD133 identify a population of CD34+ lymphatic/vascular endothelial precursor cells. *Blood, 101*, 168–172.

Sano, Y., Ksander, B. R., & Streilein, J. W. (1995). Fate of orthotopic corneal allografts in eyes that cannot support anterior chamber-associated immune deviation induction. *Investigative Ophthalmology & Visual Science, 36*, 2176–2185.

Singh, N., Tiem, M., Watkins, R., Cho, Y. K., Wang, Y., Olsen, T., et al. (2013). Soluble vascular endothelial growth factor receptor-3 is essential for corneal alymphaticity. *Blood, 121*(20), 4242–4249.

Steven, P., Bock, F., Huttmann, G., & Cursiefen, C. (2011). Intravital two-photon microscopy of immune cell dynamics in corneal lymphatic vessels. *PloS One, 6*, e26253.

Streilein, J. W., Yamada, J., Dana, M. R., & Ksander, B. R. (1999). Anterior chamber-associated immune deviation, ocular immune privilege, and orthotopic corneal allografts. *Transplantation Proceedings, 31*, 1472–1475.

Tammela, T., & Alitalo, K. (2010). Lymphangiogenesis: Molecular mechanisms and future promise. *Cell, 140*, 460–476.

Truong, T., Altiok, E., Yuen, D., Ecoiffier, T., & Chen, L. (2011). Novel characterization of lymphatic valve formation during corneal inflammation. *PloS One, 6*, e21918.

Xi, X., McMillan, D. H., Lehmann, G. M., Sime, P. J., Libby, R. T., Huxlin, K. R., et al. (2011). Ocular fibroblast diversity: Implications for inflammation and ocular wound healing. *Investigative Ophthalmology & Visual Science, 52*, 4859–4865.

Yamagami, S., & Dana, M. R. (2001). The critical role of lymph nodes in corneal alloimmunization and graft rejection. *Investigative Ophthalmology & Visual Science, 42*, 1293–1298.

Yamagami, S., Dana, M. R., & Tsuru, T. (2002). Draining lymph nodes play an essential role in alloimmunity generated in response to high-risk corneal transplantation. *Cornea, 21*, 405–409.

Yin, N., Zhang, N., Xu, J., Shi, Q., Ding, Y., & Bromberg, J. S. (2011). Targeting lymphangiogenesis after islet transplantation prolongs islet allograft survival. *Transplantation, 92*, 25–30.

Chapter 11
The Role of Neuropilin-1/Semaphorin 3A Signaling in Lymphatic Vessel Development and Maturation

Alexandra M. Ochsenbein*, Sinem Karaman*, Giorgia Jurisic, and Michael Detmar

Abstract During development, the lymphatic and the blood vascular system form highly branched networks that show extensive architectural similarities with the peripheral nervous system. Increasing evidence suggests that the vascular and the nervous systems share signaling pathways to overcome common challenges such as guidance of growth and patterning. Semaphorins, a large group of proteins originally identified as axon guidance molecules with repelling function, and their receptors, neuropilins and plexins, have recently also been implicated in vascular development. Here, we summarize the role of semaphorins and their receptors in angiogenesis and lymphangiogenesis, with an emphasis on neuropilin-1/semaphorin 3A interactions in lymphatic vessel maturation and valve formation. Understanding the basic principles of lymphatic vessel development and maturation might facilitate the development of therapies for the treatment of human diseases associated with lymphedema.

11.1 Axon Guidance Molecules in Vascular Development

During development, the lymphatic and the blood vascular system form highly branched networks that show extensive architectural similarities to the peripheral nervous system (Meadows et al. 2012). Accumulating evidence suggests that the molecular pathways used to determine the specific growth patterns and to wire the whole organism are partially shared between both systems (Carmeliet 2003).

* Equally contributing authors. The authors declare no conflicts of interest.

A.M. Ochsenbein • S. Karaman • M. Detmar (✉)
Institute of Pharmaceutical Sciences, Swiss Federal Institute of Technology, ETH Zurich, Wolfgang Pauli-Str. 10, HCI H303, 8093 Zurich, Switzerland
e-mail: michael.detmar@pharma.ethz.ch

G. Jurisic
Novartis Institutes for Biomedical Research, Basel, Switzerland

F. Kiefer and S. Schulte-Merker (eds.), *Developmental Aspects of the Lymphatic Vascular System*, Advances in Anatomy, Embryology and Cell Biology 214,
DOI 10.1007/978-3-7091-1646-3_11, © Springer-Verlag Wien 2014

Endothelial cells and neurons encounter similar challenges (path finding, patterning, alignment, etc.) in establishing their networks, which might explain why they often develop in a coordinated fashion (Carmeliet 2003). Angiogenic tip cells and axonal growth cones utilize common guidance cues to navigate through the tissues towards their targets, and both cell types possess receptors for these signaling molecules (Adams and Eichmann 2010; Larrivée et al. 2009). These include roundabout/slit (Robo/Slit), netrin–UNC5/neogenin, ephrin/ephrin-B, and the semaphorin/neuropilin/plexin (Sema/Nrp/Plxn) system (Adams and Eichmann 2010). The Sema/Nrp/Plxn system comprises highly complex signaling pathways, in which the outcome depends on combinatorial Sema/Nrp/Plxn codes, spatiotemporal expression differences, and involved cell types.

11.2 Semaphorins and Their Receptors

Semaphorins are a large family of secreted and membrane-bound proteins that share a Sema domain and that have been classified into eight classes, according to their further structural domains: invertebrate semaphorins classes 1, 2, and 5, vertebrate semaphorins classes 3–7, and the virally encoded class V semaphorins. Semaphorins of classes 1, 4, 5, 6, and 7 are membrane bound, whereas semaphorins 2, 3, and V are secreted. Membrane-bound semaphorins bind and signal directly via plexins. Secreted semaphorins bind to a holoreceptor complex of Nrp and plexins (reviewed in (Zhou et al. 2008)) with the exception of Sema3E, which is reported to directly bind plexin-D1, independent of Nrps (Gu et al. 2005). Sema3A dimers bind Nrp-1 and Sema3A downstream signaling might be enhanced by plexinA1 (Takahashi et al. 1999) and plexinD1 (Gitler et al. 2004). Plexins are single-pass transmembrane receptors, which have an extracellular Sema domain and are subdivided into four groups: A, B, C, and D (reviewed in (Hota and Buck 2012; Perala et al. 2012)). The intracellular domains of plexins have GAP activity towards the small GTPase R-Ras (Oinuma et al. 2004).

Neuropilins (Nrp-1 and Nrp-2) are single-pass transmembrane molecules with an approximate size of 100 kDa (reviewed in (Raimondi and Ruhrberg 2013)). They serve as co-receptors with plexins and vascular endothelial growth factor receptors (VEGFRs) and facilitate signal transduction of semaphorins and VEGFs during cardiovascular and nervous system development (Gu et al. 2003). Their extracellular domain consists of two complement-binding C1r/C1s, Uegf, and Bmp1 (CUB) domains—termed a1/a2 domains—that mediate protein–protein associations in the complement system. In Nrps, these are responsible for semaphorin binding. The b1/b2 domains consist of two FV/FVIII coagulation factor-like domains that are responsible for VEGF binding. The c domain that contains a meprin, A-5 protein, and receptor protein–tyrosine phosphatase mu (MAM) domain is thought to be important for Nrp oligomerization (Gu et al. 2002; Nakamura et al. 1998). The short cytoplasmic tail of neuropilin-1, which contains a PDZ (postsynaptic density protein (PSD95), Drosophila disc large tumor suppressor (DlgA), and zonula

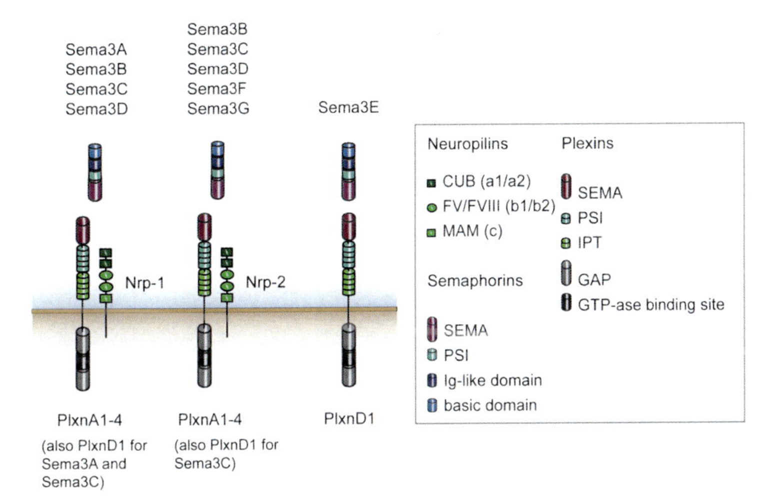

Fig. 11.1 Binding patterns of class 3 semaphorins. Class 3 semaphorins consist of SEMA, variable PSI (plexins, sema, integrins), and immunoglobulin (Ig)-like domains. Plexins contain SEMA, PSI, and IPT (Ig-like, plexins, transcription factors) domains on the extracellular and GAP and GTPase-binding domains on the intracellular side. Nrps have a very short intracellular domain. The extracellular part of Nrps consists of complement-binding domains (CUB) a1 and a2, which bind to semaphorins, coagulation factor homology domains (FV/FIII) b1 and b2, which bind VEGFs, and a MAM domain. Sema3A binds Nrp-1. Sema3F and Sema3G bind Nrp-2. Sema3B, Sema3C, and Sema3D bind both Nrps. Sema3E signals directly via plexinD1

occludens-1 protein) domain, binds synectin and is also involved in the stable complex formation with VEGFR-2 (Fig. 11.1) (Prahst et al. 2008).

Semaphorins and VEGFs bind Nrp-1 and Nrp-2 with varying and partly overlapping specificities, but they do not compete for Nrp binding (Appleton et al. 2007). Neuropilin-1 binds Sema3A, Sema3B, Sema3C, and Sema3D, whereas Nrp-2 binds Sema3B, Sema3C, Sema3D, Sema3F, and Sema3G (Fig. 11.1) (Chen et al. 1997; Taniguchi et al. 2005). Of the VEGFs, Nrp-1 binds VEGF-A (isoforms 121 and 165), VEGF-B, VEGF-C, VEGF-D, and VEGF-E. Nrp-2 binds VEGF-C, VEGF-D, and VEGF-A_{165} and VEGF-A_{145} (reviewed in (Pellet-Many et al. 2008)).

11.3 Neuropilins in Vascular Development

Neuropilins comprise versatile molecules that convey different signals from structurally distinct ligands. Nrp-1 was first identified in the central nervous system as a receptor involved in axon guidance and in vitro growth cone collapse (Takagi et al. 1987). Recent studies point to a further role of Nrp-1 signaling in vascular developmental processes by binding VEGFs via the b1/b2 domains (Karpanen et al. 2006). Nrp-1-deficient mice showed extensive cardiovascular defects such as impaired tip cell guidance, disorganized vessels, and a sparse capillary network

Table 11.1 Overview of genetic mouse models involving the Nrp/Sema axis

Gene	Model	Lethality	Vascular phenotype	References
Nrp-1	KO	E12.5	Severe vascular abnormalities	Kitsukawa et al. (1995), Kawasaki et al. (1999)
Nrp-1	Sema$^{-/-}$ (mutant lacking Sema-binding domain)	40 % until P7	No vascular phenotype reported, growth retarded	Gu et al. (2003)
Nrp-1	C/C; Cre (endothelial cell specific KO, CreTie2)	Perinatal	Multiple cardiac defects: persistent truncus arteriosus, anomalous origin of the coronary artery, lack of septation of cardiac outflow	Gu et al. (2003)
Nrp-2	KO	Viable and fertile	Defects in lymphatic capillary formation	Chen et al. (2000), Yuan et al. (2002)
Sema3A	KO	Variable	Depending on the background defective or normal vascular development	Behar et al. (1996), Taniguchi et al. (2005), Serini et al. (2003), Vieira et al. (2007)
Sema3C	KO	Postnatal mortality: 50 % on C57BL/6 and 129, 96 % on CD1 background	Persistent truncus arteriosus, interruption of the aortic arch, abnormal migration of cardial crest cells	Feiner et al. (2001)
Sema3E	KO	Viable and fertile	Disorganized intersomitic vessels	Gu et al. (2005)

KO knockout

(Table 11.1) (Kawasaki et al. 1999; Kitsukawa et al. 1995). Nrp-1 overexpression in a chimeric mouse model also led to cardiovascular and skeletal malformations and conversely resulted in increased capillary growth (Table 11.1) (Kitsukawa et al. 1995). Similarly, specific deletion of Nrp-1 in endothelial cells resulted in multiple cardiac defects (Gu et al. 2003). During early vascular development, Nrp-1 signaling is apparently essential for the guidance of endothelial tip cell filopodia in the hindbrain (Gerhardt et al. 2004). Nrp-1 signaling is also important in vascular remodeling in the postnatal mouse retina, as shown by the use of function-blocking antibodies (Pan et al. 2007).

In contrast to Nrp-1, which is mainly expressed on arteries, Nrp-2 is predominantly expressed on veins and lymphatic vessels (Herzog et al. 2001). Nrp-2-deficient mice survived until adulthood; however, they were reported to have smaller and fewer lymphatic vessels and to show defects in capillary formation (Table 11.1) (Chen et al. 2000; Yuan et al. 2002; Giger et al. 2000). Nrp-2 seems to be involved in VEGF-C-mediated lymphatic vascular sprouting, together with VEGFR-3 (Xu et al. 2010). A double KO of Nrp-1 and Nrp-2 resulted in a more severe phenotype with lethality at embryonic day (E) 8, similar to VEGF-A_{165} and VEGFR-2 KO mice (Giger et al. 2000).

11.4 Semaphorins in Blood Vascular Development

Apart from their role in axon guidance, semaphorins, in particular class 3 semaphorins, are also involved in developmental and in pathological angiogenesis in the adult, including cancer (reviewed in (Gu and Giraudo 2013)). Sema3E-plexinD1 signaling might balance the tip/stalk cell ratio and diminish blood vessel branching in murine neonatal retinas (Kim et al. 2011). In Sema3E KO mice, the vasculogenesis process was disturbed, leading to abnormally shaped, paired dorsal aortae and numerous ectopic vessels that extended into the lateral avascular regions (Table 11.1) (Meadows et al. 2012). Sema3C mutant mice died within hours after birth from cardiovascular defects. These consisted of interruption of the aortic arch and improper separation of the cardiac outflow tract (Feiner et al. 2001). Sema3F induced cytoskeletal collapse in human umbilical vein endothelial cells (HUVECs) and inhibited cell contractility and cell migration (Shimizu et al. 2008). In vitro, anti-angiogenic Sema3F effects were synergistically enhanced by Sema3A (Guttmann-Raviv et al. 2007), and Sema3A inhibited endothelial cell migration by inhibiting integrin function (Serini et al. 2003). At high concentrations, Sema3A compromised HUVEC survival (Guttmann-Raviv et al. 2007).

In vivo, the role of Sema3A in vascular development is complex. In quail embryos, ectopic expression of Sema3A decreased the vascular capillary density, whereas application of a Sema3A antibody at E4.5 led to hypervascularization (Bates et al. 2003). Vascular defects were observed in E9.5 Sema3A$^{-/-}$ mouse embryos, including a lack of anterior cardinal vein branching (Serini et al. 2003), whereas the vasculature of the hindbrains appeared normal (Vieira et al. 2007). Nrp-1^{sema} mice, which have a KO of the Sema-binding domain of Nrp-1, abolishing the Sema-induced, but not the VEGF-induced Nrp-1 signaling, showed no vascular phenotype at E12.5, whereas Tie2-Cre Nrp-1 KO mice had dramatic systemic vascular deficiencies (Table 11.1) (Gu et al. 2003), suggesting that Sema3A-induced Nrp-1 signaling in endothelial cells has no general effect on angiogenesis. However, Nrp-1^{sema} mice showed a bilateral atrial enlargement—a defect also noted in the Sema3A null mice (Gu et al. 2003).

11.5 Role of Neuropilin-1/Semaphorin 3A Signaling in Lymphatic Vascular Development

Using high-speed cell sorting of lymphatic and blood vascular endothelial cells directly from mouse tissue, followed by gene expression profiling by microarray analysis, our laboratory recently identified previously unreported molecules that are specifically expressed in lymphatic or blood vascular endothelium (Jurisic et al. 2012). We found that Sema3A was specifically expressed in lymphatic vessels, whereas its receptor Nrp-1 was highly expressed on blood vessels and by perivascular cells (Jurisic et al. 2012). In utero treatment of mice with an antibody that blocks Sema3A, but not VEGF-A binding to Nrp-1, resulted in a reduced draining function of lymphatic vessels, an abnormal structure of collecting lymphatic vessels and lymphatic valves, and an aberrant smooth muscle cell coverage of lymphatic vessels (Jurisic et al. 2012). Since Nrp-1 was found to be expressed by blood vascular endothelial cells, perivascular smooth muscle cells, and lymphatic valve cells and since Sema3A is produced by the lymphatic endothelial cells, it is conceivable that Sema3A is involved in blood vessel/lymphatic vessel separation by transducing repulsive signals through Nrp-1 to blood vascular endothelial cells (Fig. 11.2). Moreover, the characteristic sparse coverage of lymphatic vessels by perivascular cells might be maintained via the production of Sema3A by lymphatic endothelium, in agreement with the inhibitory effect of Sema3A on pericyte migration in vitro (Jurisic et al. 2012). Nrp-1 expression by lymphatic valve cells might contribute to keep the valve cells separated from the lymphatic vessel wall.

These concepts are supported by findings that Sema3A mRNA was expressed by postnatal day (P) 0 mesenteric lymphatic vessels, but not by mesenteric arteries or veins, and that alkaline-phosphatase (AP)-tagged Sema3A protein bound not only to Nrp-1 expressing arteries and veins but also to lymphatic valves (Bouvree et al. 2012). Nrp-1 and plexinA1 mRNA were detected in the valve-forming areas, suggesting that Sema3A/Nrp-1/plexinA1 signaling is involved in lymphatic valve formation. Indeed, Sema3A$^{-/-}$, Nrp-1^{sema} KO, and plexinA1-deficient mice all had smaller mesenteric lymphatic valve areas. Abnormal smooth muscle cell coverage of mesenteric valve regions was seen in Sema3A$^{-/-}$ mice and after blockade of Sema3A/Nrp-1 binding by antibodies. Defective lymphatic valve formation is known to result in impaired lymphatic function (Sabine et al. 2012; Petrova et al. 2004). It is of interest that Sema3A$^{-/-}$ mice, as well as mice treated with Sema-binding blocking antibody, had defects in collecting lymphatic valves, which were insufficient to prevent lymphatic backflow, indicating impaired lymphatic drainage.

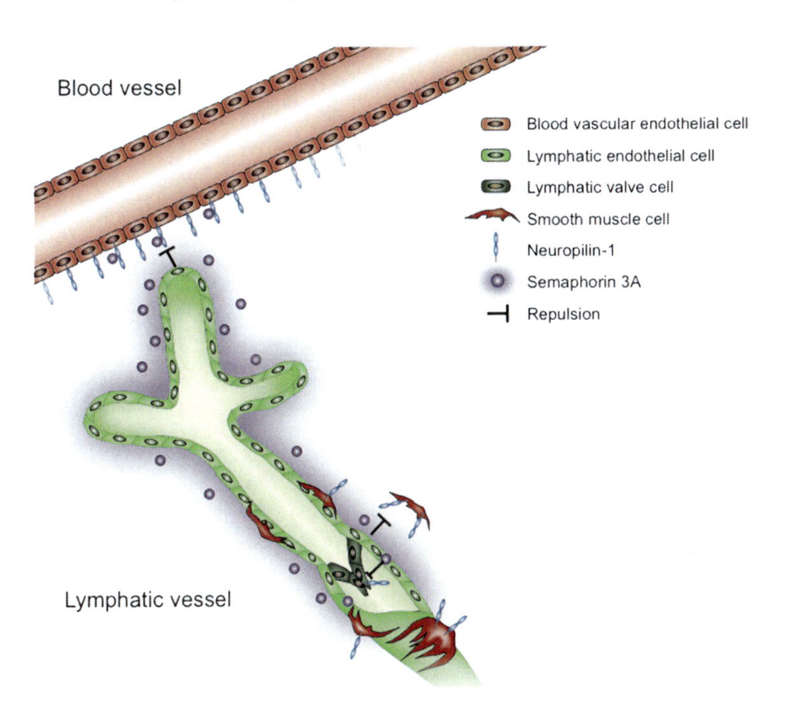

Fig. 11.2 Model of potential Sema3A/Nrp-1 signaling loops with impact on the lymphatic vasculature. Nrp-1 is expressed by blood vascular endothelial cells, smooth muscle cells, and lymphatic valve cells. Sema3A, produced by lymphatic endothelial cells, might play a role in blood vessel/lymphatic vessel separation by transducing repulsive signals through Nrp-1 on blood vascular endothelial cells. Lymphatic endothelium maintains a sparse coverage of Nrp-1 expressing smooth muscle cells by conveying repulsive signals via the production of Sema3A. Enhanced Nrp-1 expression on lymphatic valve cells might help to keep the valve cells separate from the lymphatic vessel wall

11.6 Concluding Remarks and Future Directions

In addition to the established function of neuropilin/semaphorin signaling in axonal guidance, class 3 semaphorins and their receptors also play important roles in lymphatic and blood vascular development. The differential expression patterns of semaphorins and their receptors, e.g., the expression of Nrp-1 on arteries, veins, perivascular cells, and lymphatic valves and of Sema3A by lymphatic vessels, suggest spatially controlled functions. However, the detailed molecular and cellular interactions that guide lymphatic vessel development, maturation, and function still need to be investigated. Moreover, several questions remain to be addressed:

Which cells are the main producers of Sema3A and what are its functions during blood vascular and lymphatic vascular development? What are the functions of endothelial cell-derived and in particular lymphatic endothelium-derived Sema3A in cancer, inflammation, and immune tolerance development? The early embryonic lethality of systemic Nrp-1 KO mice and the early postnatal lethality of systemic Sema3A KO mice in the 129 genetic background highlight the need for more sophisticated experimental models to investigate these questions. Future work should therefore utilize novel genetic approaches, such as targeted Cre-LoxP systems, to shed more light onto the functional role of Nrp-1/Sema3A interactions during lymphatic vascular development.

Acknowledgements Work in the authors' laboratory has been supported by National Institutes of Health grant CA69184, Swiss National Science Foundation grants 3100A0_108207, 31003A_130627, and 310030B_147087, Commission of the European Communities grant LSHC-CT-2005-518178, Advanced European Research Council grant LYVICAM, the Leducq Foundation Transatlantic Network of Excellence grant Lymph Vessels in Obesity and Cardiovascular Disease, Oncosuisse, and Krebsliga Zurich.

References

Adams, R. H., & Eichmann, A. (2010). Axon guidance molecules in vascular patterning. *Cold Spring Harbor Perspectives in Biology, 2*(5), a001875.

Appleton, B. A., Wu, P., Maloney, J., Yin, J., Liang, W.-C., Stawicki, S., et al. (2007). Structural studies of neuropilin/antibody complexes provide insights into semaphorin and VEGF binding. *EMBO Journal, 26*(23), 4902–4912.

Bates, D., Taylor, G. I., Minichiello, J., Farlie, P., Cichowitz, A., Watson, N., et al. (2003). Neurovascular congruence results from a shared patterning mechanism that utilizes Semaphorin3A and Neuropilin-1. *Developmental Biology, 255*(1), 77–98.

Behar, O., Golden, J. A., Mashimo, H., Schoen, F. J., Fishman, M. C. (1996). Semaphorin III is needed for normal patterning and growth of nerves, bones and heart. *Nature, 383*(6600), 525–528.

Bouvree, K., Brunet, I., Del Toro, R., Gordon, E., Prahst, C., Cristofaro, B., et al. (2012). Semaphorin3A, neuropilin-1, and plexinA1 are required for lymphatic valve formation. *Circulation Research, 111*(4), 437–445.

Carmeliet, P. (2003). Blood vessels and nerves: Common signals, pathways and diseases. *Nature Reviews Genetics, 4*(9), 710–720.

Chen, H., Bagri, A., Zupicich, J. A., Zou, Y., Stoeckli, E., Pleasure, S. J., et al. (2000). Neuropilin-2 regulates the development of selective cranial and sensory nerves and hippocampal mossy fiber projections. *Neuron, 25*(1), 43–56.

Chen, H., Chédotal, A., He, Z., Goodman, C. S., & Tessier-Lavigne, M. (1997). Neuropilin-2, a novel member of the neuropilin family, is a high affinity receptor for the semaphorins Sema E and Sema IV but not Sema III. *Neuron, 19*(3), 547–559.

Feiner, L., Webber, A. L., Brown, C. B., Lu, M. M., Jia, L., Feinstein, P., et al. (2001). Targeted disruption of semaphorin 3C leads to persistent truncus arteriosus and aortic arch interruption. *Development, 128*(16), 3061–3070.

Gerhardt, H., Ruhrberg, C., Abramsson, A., Fujisawa, H., Shima, D., & Betsholtz, C. (2004). Neuropilin-1 is required for endothelial tip cell guidance in the developing central nervous system. *Developmental Dynamics, 231*(3), 503–509.

Giger, R. J., Cloutier, J. F., Sahay, A., Prinjha, R. K., Levengood, D. V., Moore, S. E., et al. (2000). Neuropilin-2 is required in vivo for selective axon guidance responses to secreted semaphorins. *Neuron, 25*(1), 29–41.

Gitler, A. D., Lu, M. M., & Epstein, J. A. (2004). PlexinD1 and semaphorin signaling are required in endothelial cells for cardiovascular development. *Developmental Cell, 7*(1), 107–116.

Gu, C., & Giraudo, E. (2013). The role of semaphorins and their receptors in vascular development and cancer. *Experimental Cell Research, 319*(9), 1306–1316.

Gu, C., Limberg, B. J., Whitaker, G. B., Perman, B., Leahy, D. J., Rosenbaum, J. S., et al. (2002). Characterization of neuropilin-1 structural features that confer binding to semaphorin 3A and vascular endothelial growth factor 165. *Journal of Biological Chemistry, 277*(20), 18069–18076.

Gu, C., Rodriguez, E. R., Reimert, D. V., Shu, T., Fritzsch, B., Richards, L. J., et al. (2003). Neuropilin-1 conveys semaphorin and VEGF signaling during neural and cardiovascular development. *Developmental Cell, 5*(1), 45–57.

Gu, C., Yoshida, Y., Livet, J., Reimert, D. V., Mann, F., Merte, J., et al. (2005). Semaphorin 3E and plexin-D1 control vascular pattern independently of neuropilins. *Science, 307*(5707), 265–268.

Guttmann-Raviv, N., Shraga-Heled, N., Varshavsky, A., Guimaraes-Sternberg, C., Kessler, O., & Neufeld, G. (2007). Semaphorin-3A and semaphorin-3F work together to repel endothelial cells and to inhibit their survival by induction of apoptosis. *Journal of Biological Chemistry, 282*(36), 26294–26305.

Herzog, Y., Kalcheim, C., Kahane, N., Reshef, R., & Neufeld, G. (2001). Differential expression of neuropilin-1 and neuropilin-2 in arteries and veins. *Mechanisms of Development, 109*(1), 115–119.

Hota, P. K., & Buck, M. (2012). Plexin structures are coming: Opportunities for multilevel investigations of semaphorin guidance receptors, their cell signaling mechanisms, and functions. *Cellular and Molecular Life Sciences, 69*(22), 3765–3805.

Jurisic, G., Maby-El Hajjami, H., Karaman, S., Ochsenbein, A. M., Alitalo, A., Siddiqui, S. S., et al. (2012). An unexpected role of semaphorin3A/neuropilin-1 signaling in lymphatic vessel maturation and valve formation. *Circulation Research, 111*(4), 426–436.

Karpanen, T., Heckman, C. A., Keskitalo, S., Jeltsch, M., Ollila, H., Neufeld, G., et al. (2006). Functional interaction of VEGF-C and VEGF-D with neuropilin receptors. *FASEB Journal, 20* (9), 1462–1472.

Kawasaki, T., Kitsukawa, T., Bekku, Y., Matsuda, Y., Sanbo, M., Yagi, T., et al. (1999). A requirement for neuropilin-1 in embryonic vessel formation. *Development, 126*(21), 4895–4902.

Kim, J., Oh, W.-J., Gaiano, N., Yoshida, Y., & Gu, C. (2011). Semaphorin 3E-Plexin-D1 signaling regulates VEGF function in developmental angiogenesis via a feedback mechanism. *Genes and Development, 25*(13), 1399–1411.

Kitsukawa, T., Shimono, A., Kawakami, A., Kondoh, H., & Fujisawa, H. (1995). Overexpression of a membrane protein, neuropilin, in chimeric mice causes anomalies in the cardiovascular system, nervous system and limbs. *Development, 121*(12), 4309–4318.

Larrivée, B., Freitas, C., Suchting, S., Brunet, I., & Eichmann, A. (2009). Guidance of vascular development: Lessons from the nervous system. *Circulation Research, 104*(4), 428–441.

Meadows, S. M., Fletcher, P. J., Moran, C., Xu, K., Neufeld, G., Chauvet, S., et al. (2012). Integration of repulsive guidance cues generates avascular zones that shape mammalian blood vessels. *Circulation Research, 110*(1), 34–46.

Nakamura, F., Tanaka, M., Takahashi, T., Kalb, R. G., & Strittmatter, S. M. (1998). Neuropilin-1 extracellular domains mediate semaphorin D/III-induced growth cone collapse. *Neuron, 21*(5), 1093–1100.

Oinuma, I., Ishikawa, Y., Katoh, H., & Negishi, M. (2004). The Semaphorin 4D receptor Plexin-B1 is a GTPase activating protein for R-Ras. *Science, 305*(5685), 862–865.

Pan, Q., Chanthery, Y., Liang, W.-C., Stawicki, S., Mak, J., Rathore, N., et al. (2007). Blocking neuropilin-1 function has an additive effect with anti-VEGF to inhibit tumor growth. *Cancer Cell, 11*(1), 53–67.

Pellet-Many, C., Frankel, P., Jia, H., & Zachary, I. (2008). Neuropilins: Structure, function and role in disease. *Biochemical Journal, 411*(2), 211–226.

Perala, N., Sariola, H., & Immonen, T. (2012). More than nervous: The emerging roles of plexins. *Differentiation, 83*(1), 77–91.

Petrova, T. V., Karpanen, T., Norrmen, C., Mellor, R., Tamakoshi, T., Finegold, D., et al. (2004). Defective valves and abnormal mural cell recruitment underlie lymphatic vascular failure in lymphedema distichiasis. *Nature Medicine, 10*(9), 974–981.

Prahst, C., Héroult, M., Lanahan, A. A., Uziel, N., Kessler, O., Shraga-Heled, N., et al. (2008). Neuropilin-1-VEGFR-2 complexing requires the PDZ-binding domain of neuropilin-1. *Journal of Biological Chemistry, 283*(37), 25110–25114.

Raimondi, C., & Ruhrberg, C. (2013). Neuropilin signalling in vessels, neurons and tumours. *Seminars in Cell and Developmental Biology, 24*(3), 172–178.

Sabine, A., Agalarov, Y., Maby-El Hajjami, H., Jaquet, M., Hagerling, R., Pollmann, C., et al. (2012). Mechanotransduction, PROX1, and FOXC2 cooperate to control connexin37 and calcineurin during lymphatic-valve formation. *Developmental Cell, 22*(2), 430–445.

Serini, G., Valdembri, D., Zanivan, S., Morterra, G., Burkhardt, C., Caccavari, F., et al. (2003). Class 3 semaphorins control vascular morphogenesis by inhibiting integrin function. *Nature, 424*(6947), 391–397.

Shimizu, A., Mammoto, A., Italiano, J. E., Jr., Pravda, E., Dudley, A. C., Ingber, D. E., et al. (2008). ABL2/ARG tyrosine kinase mediates SEMA3F-induced RhoA inactivation and cytoskeleton collapse in human glioma cells. *Journal of Biological Chemistry, 283*(40), 27230–27238.

Takagi, S., Tsuji, T., Amagai, T., Takamatsu, T., & Fujisawa, H. (1987). Specific cell surface labels in the visual centers of *Xenopus laevis* tadpole identified using monoclonal antibodies. *Developmental Biology, 122*(1), 90–100.

Takahashi, T., Fournier, A., Nakamura, F., Wang, L. H., Murakami, Y., Kalb, R. G., et al. (1999). Plexin-neuropilin-1 complexes form functional semaphorin-3A receptors. *Cell, 99*(1), 59–69.

Taniguchi, M., Masuda, T., Fukaya, M., Kataoka, H., Mishina, M., Yaginuma, H., et al. (2005). Identification and characterization of a novel member of murine semaphorin family. *Genes to Cells, 10*(8), 785–792.

Vieira, J. M., Schwarz, Q., & Ruhrberg, C. (2007). Selective requirements for NRP1 ligands during neurovascular patterning. *Development, 134*(10), 1833–1843.

Xu, Y., Yuan, L., Mak, J., Pardanaud, L., Caunt, M., Kasman, I., et al. (2010). Neuropilin-2 mediates VEGF-C-induced lymphatic sprouting together with VEGFR3. *Journal of Cell Biology, 188*(1), 115–130.

Yuan, L., Moyon, D., Pardanaud, L., Breant, C., Karkkainen, M. J., Alitalo, K., et al. (2002). Abnormal lymphatic vessel development in neuropilin 2 mutant mice. *Development, 129*(20), 4797–4806.

Zhou, Y., Gunput, R. A., & Pasterkamp, R. J. (2008). Semaphorin signaling: Progress made and promises ahead. *Trends in Biochemical Sciences, 33*(4), 161–170.

Chapter 12
A Fisheye View on Lymphangiogenesis

Andreas van Impel and Stefan Schulte-Merker

Abstract Zebrafish have been widely used to study vasculogenesis and angiogenesis, and the vascular system is one of the most intensively studied organ systems in teleosts. It is a little surprising, therefore, that the development of the zebrafish lymphatic network has only been investigated in any detail for less than a decade now. In those last few years, however, significant progress has been made. Due to favorable imaging possibilities within the early zebrafish embryo, we have a very good understanding of what cellular behavior accompanies the formation of the lymphatic system and which cells within the vasculature are destined to contribute to lymphatic vessels. The migration routes of future lymphatic endothelial cells have been monitored in great detail, and a number of transgenic lines have been developed that help to distinguish between arterial, venous, and lymphatic fates in vivo. Furthermore, both forward and reverse genetic tools have been systematically employed to unravel which genes are involved in the process. Not surprisingly, a number of known players were identified (such as *vegfc* and *flt4*), but work on zebrafish has also distinguished genes and proteins that had not previously been connected to lymphangiogenesis. Here, we will review these topics and also compare the equivalent stages of lymphatic development in zebrafish and mice. We will, in addition, highlight some of those studies in zebrafish that have helped to identify and to further characterize human disease conditions.

Research on the lymphatic system in teleosts has taken a somewhat convoluted path through the years: the first description can be found as early as 1769 by Hewson and Hunter (1769), but in the last century the notion prevailed that fish do not have lymphatics (Vogel and Claviez 1981), and there was certainly very little work done on them. Only with the description of embryonic lymphatic vessels, paired with the

A. van Impel • S. Schulte-Merker (✉)
Hubrecht Institute, KNAW and UMC Utrecht, Uppsalalaan 8, 3584 CT Utrecht,
The Netherlands
e-mail: s.schulte@hubrecht.eu

F. Kiefer and S. Schulte-Merker (eds.), *Developmental Aspects of the Lymphatic Vascular System*, Advances in Anatomy, Embryology and Cell Biology 214,
DOI 10.1007/978-3-7091-1646-3_12,

demonstration that these vessels can take up substances from the interstitium and are dependent on *vegfc* function (Küchler et al. 2006; Yaniv et al. 2006), it has been fully appreciated that lymphatic vessels exist in fish. Very quickly, the existence of transgenic lines and the favorable imaging opportunities in zebrafish larvae have led to a detailed understanding of how lymphangiogenesis occurs in early teleost embryos, and functional studies, based on morpholino approaches or on dedicated mutant screens, led to the identification of key genes in the process. In this review, we would like to discuss our current knowledge, with a particular focus on the comparison between lymphangiogenesis in the zebrafish and the mouse embryo.

12.1 Venous Spouting and the Origin of Lymphatics in Fish

As in amphibians (Ny et al. 2005) and amniotes (Chap. 2), the lymphatic system in teleosts stems from the venous system, but in contrast to mice, its initial development occurs simultaneously with the venous system. Hence, initial lymphangiogenesis cannot be understood without also considering arterial and venous angiogenesis.

After the formation of a dorsal aorta (DA) and a posterior cardinal vein (PCV) in the midline of the embryo through the process of vasculogenesis, an initial wave of angiogenesis forms about 30 pairs of intersegmental vessels (ISVs) that all sprout from the DA (Fig. 12.1a, leftmost diagram) and are therefore, at this point in time, referred to as intersegmental arteries. At around 36 hpf, a second wave of angiogenesis occurs, this time exclusively from the PCV. Again, there are 30 pairs of bilateral sprouts, but in the case of these venous sprouts, there are two different kinds of behavior and cellular fates to be observed: about half of the venous sprouts very quickly form a stable and eventually lumenized connection with one of the nearby arterial ISVs. This ISV has, for a short while, a connection with both the DA and the PCV (Fig. 12.1a, 40 hpf diagram), but the connection to the DA gets lost quickly, leaving only the connection to the PCV and completing the process of remodeling an artery into a venous intersegmental vessel (vISV; Fig. 12.1a, 54 hpf diagram). Which specific factors determine whether a particular venous sprout will connect to an arterial ISV (or not) is unclear at present; suppressing the function of delta-like-4, Notch-1b, or Notch-6 will result in the almost exclusive formation of veins (Geudens et al. 2010), suggesting a role of the Delta/Notch pathway in the process. However, there is no clear demonstration of either of these genes being expressed in the PCV or in venous sprouts; hence more work is required here to understand this aspect of vein formation in zebrafish.

In consequence, half of the venous sprouts turn 30 aISVs into 30 vISVs in the wild-type situation. What about the other half of the venous sprouts?

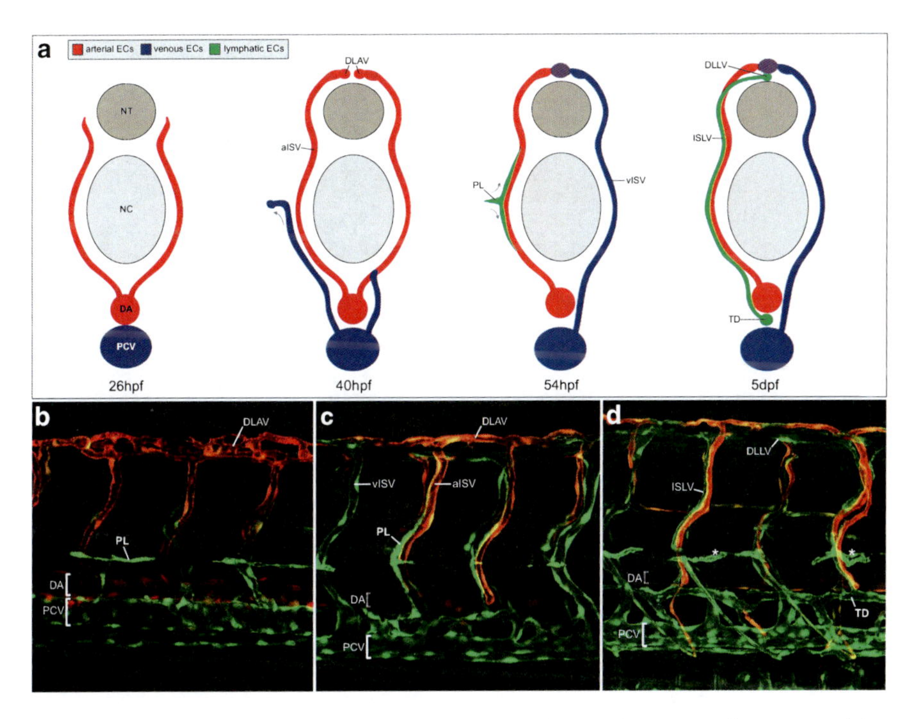

Fig. 12.1 Lymphangiogenesis in the zebrafish trunk. (**a**) Schematic cross-sections summarizing the main events during the formation of the trunk vasculature. During primary sprouting, a set of arterial intersegmental vessels (aISVs) is formed by sprouting from the dorsal aorta (DA; see 26 hpf). These angiogenic sprouts migrate to the dorsal-most aspect of the embryo where they T-branch, giving rise to an initial pair of dorsal longitudinal anastomotic vessels (DLAV), which later fuse. In a second step, sprouts from the posterior cardinal vein (PCV) emerge at around 36 hpf. Approximately half of these sprouts migrate dorsally and connect to a preexisting aISV, thereby remodeling an artery into an intersegmental vein (vISV; see 40 hpf *right sprout*). The other half of the venous sprouts migrates further dorsal to the midline of the embryo where they turn towards the so-called horizontal myoseptum region at the surface of the embryo (see 40 hpf *left sprout*). This second group of ECs will give rise to lymphatic precursors which are referred to as parachordal lymphangioblasts (PLs) as soon as they lose their connection to the PCV. The PL cells will subsequently migrate exclusively along intersegmental arteries (aISVs) in a dorsal or ventral direction (see 54 hpf) to form the primitive lymphatic vasculature of the zebrafish embryo consisting of the thoracic duct (TD), positioned between the DA and the PCV, the intersegmental lymphatic vessels (ISLV), as well as the dorsal longitudinal lymphatic vessel (DLLV) (see 5 dpf). (**b–d**) Lateral views of the trunk in *flt1*^{enh}*:tdTomato*; *flt4:YFP* double transgenic embryos highlighting arterial ECs in red and venous and lymphatic ECs in green. (**b**) At 42 hpf the first lymphatic sprouts are visible at the horizontal myoseptum where they align in an anterior/posterior orientation. (**c**) Embryo at 56 hpf, in which the PLs spread out in the trunk using aISVs as migration routes. (**d**) The formation of a functional lymphatic vasculature is complete at 5 dpf. Note that the cells highlighted by an *asterisk* in (**d**) are not PLs that remained at the horizontal myoseptum but reflect a different group of venous ECs that migrates in to form blood vessels at the surface of the embryo at this stage

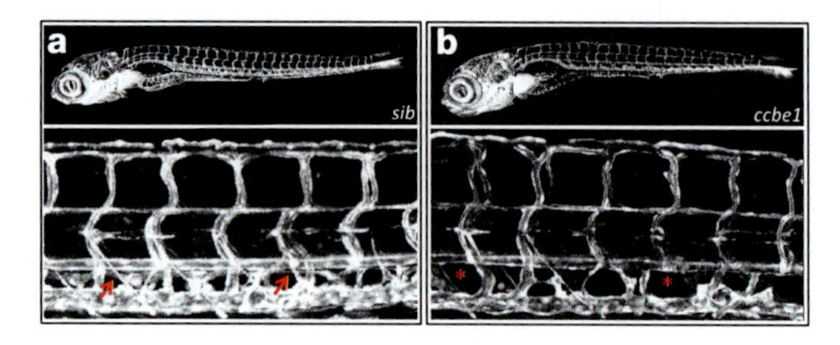

Fig. 12.2 *Ccbe1* mutant embryos lack lymphatic structures. Wild type (**a**) and *ccbe1* mutant embryo (**b**) expressing *fli1a:GFP* in all endothelial cells. At 5 dpf, *ccbe1* mutants lack the thoracic duct between DA and PCV (marked by *asterisks* in (**b**)) while retaining an overall normal blood vasculature. Note that the general appearance of the mutants is unaltered at this time point, since the formation of lymphedema only starts around 5–6 dpf

12.2 Parachordal Lymphangioblasts on the Move

The other 30 venous sprouts ignore the aISVs, and even though there is no apparent reason (such as timing or distance from the aISVs) why they should behave in a different manner, they display a completely distinct behavior. They migrate towards the dorsal midline, up to the level of the dorsal myoseptum. They then turn distally (indicated by the arrow in Fig. 12.1a, 40 hpf diagram) and migrate to the position of the horizontal myoseptum. At this point in time (around 52 hpf), these cells constitute a population of single cells that have lost the connection to the PCV. These cells have been termed parachordal lymphangioblasts (PLs) (Hogan et al. 2009a) due to the fact that they constitute the pool of cells that give rise to lymphatic endothelial cells (LECs) in the trunk and the fact that, at this point in time, they are positioned at the level of the notochord (Fig. 12.1a, 54 hpf diagram; Fig. 12.1b). Since the PLs often appear as a string of endothelial cells at the midline, they were initially described as a vessel, but it has become clear by now that they are individual cells that are not connected to each other.

The PLs are positioned for a little while at the level of the horizontal myoseptum and display migratory behavior in this region. However, they eventually move away from the midline and migrate either dorsally or ventrally (Fig. 12.1a, arrows in 54 hpf diagram; Fig. 12.1c). Intriguingly, they almost exclusively use arterial ISVs as a "substrate" to migrate on (Bussmann et al. 2010; Geudens et al. 2010). The molecular basis for this has been elucidated by Cha et al. (2012), who have demonstrated that *cxcl12b* ligand expression in arterial ECs can guide migrating PLs, which express the chemokine receptors *cxcr4a* and *cxcr4b*.

PLs that migrate in a ventral direction extend past the DA and T-branch once they reach a position in between the DA and PCV. Sprouts from two different cells, which can be a few segments apart, connect with each other and form a continuous structure that will eventually, by 5 dpf, differentiate in a fully patent thoracic duct (TD; Fig. 12.1d). PLs that migrate dorsally behave in an equivalent way and T-branch and fuse as soon as they reach the dorsal aspect of the embryo. Also at the dorsal side a continuous lymphatic vessel is formed, the so-called dorsal longitudinal lymphatic vessel (DLLV), that is connected with the thoracic duct via a set of intersegmental lymphatic vessels (ISLVs) which remain in close proximity to the arterial ISVs (Fig. 12.1a, 5 dpf diagram, and Fig. 12.1d).

12.3 Facial Lymphatics and Lymphatics in Other Parts of the Body

Other than the lymphatic structures of the trunk, there are of course also lymphatics in other parts of the body. Recently, Okuda et al. (2012) have described how, at 36 hpf, anterior lymphatics sprout from the region of the common cardinal vein and how an anterior plexus of facial lymphatics is formed during early embryogenesis. Facial lymphatics are influenced by the same genetic factors that also control lymphangiogenesis in the trunk and will not form in the absence of *vegfc* and *ccbe1*.

Furthermore, Jensen Dahl Ejby et al. (2009) describe the existence of a lymphatic system in the tail fin of teleosts. Remarkably, the study claims that in zebrafish and glass catfish these lymphatic vessels can become perfused and connected to the blood vasculature through shunts. However, whether these vessels are indeed lymphatic vessels or rather constitute a "secondary vessel system" as postulated by Vogel and colleagues (Vogel and Claviez 1981; Steffensen et al. 1986) remains to be seen.

Comparatively little attention has been paid to the adult lymphatic system. Valves have not been reported on, and there also seem to be no lymph hearts like the ones described in anuran (Kampmeier 1925) or avian embryos (Valasek et al. 2007). The interaction of LECs with support cells such as smooth muscle cells in late larval and adult stages of teleost needs to be elucidated, and more work needs to be carried out in these areas.

12.4 VEGF-C and Flt4

The secreted vascular endothelial growth factor C (VEGF-C) and its receptor, VEGFR-3/FLT4, have been shown in numerous studies to be key drivers of lymphangiogenesis. *Vegf-c* is expressed in the jugular region of early mouse embryos, at positions where migrating LECs will form the first lymphatic structures. *Vegf-c* is required for sprouting of Prox1-positive cells from the cardinal vein, and even in heterozygous situations phenotypes can be observed (Karkkainen et al. 2004; Hägerling et al. 2013). Forced expression of *Vegf-c* will induce lymphangiogenesis in transgenic mice (Jeltsch et al. 1997) and also elicits a response from venous vessels in embryonic zebrafish (Gordon et al. 2013). Proteolytic cleavage of VEGF-C and the related VEGF-D is a key step for receptor binding (Joukov et al. 1997; Stacker et al. 1999), and partial processing has an influence on which member of the VEGF receptor family will be activated upon binding.

The receptor for VEGF-C is the transmembrane receptor tyrosine kinase VEGFR-3 (also called FLT4). It is expressed on lymphatic and angiogenic blood endothelial cells (Tammela et al., 2008) together with its non-signaling co-receptor neuropilin 2 (Karpanen et al. 2006). Targeted inactivation of VEGFR-3 leads to embryonic lethality due to cardiovascular failure at E9.5, even before the formation of lymphatic vessels (Dumont et al., 1998). The phenotype is more severe than the combined depletion of VEGF-C and VEGF-D, possibly suggesting a ligand-independent function of VEGFR-3 (Haiko et al. 2008).

In zebrafish, mutants for both *vegfc* (Villefranc et al. 2013) and *flt4* (Hogan et al. 2009b) have been described, with both genes having an essential role in venous sprouting and lymphangiogenesis: in the absence of Vegfc or Flt4 activity, sprouting of both future venous and lymphatic endothelial cells is severely impeded as one might expect on the basis of both these cell populations sprouting simultaneously from the PCV. Similar to the murine situation (Zhang et al., 2010), *flt4* mutants in zebrafish that harbor a kinase activity-deficient allele (Hogan et al., 2009b) show no other cardiovascular abnormalities and can even survive to adulthood (van Impel and Schulte-Merker, unpublished).

Zebrafish *flt4* is expressed predominantly by venous (Covassin et al. 2006; Bussmann et al. 2007) and lymphatic ECs (Fig. 12.1d), while *vegfc* mRNA can be found in the DA, in ISVs, and in the hypochord. Expression of *vegfc* is dynamic, and more work is required to elucidate which one of these expression domains is relevant to which step in the genesis of venous and lymphatic structures. In the absence of anti-Vegfc antibodies that would allow an appreciation of the distribution of biologically active protein in situ, looking at the sites of mRNA production might also be somewhat misleading: the processing of the pre-pro-form of Vegfc might well be controlled locally, creating higher amounts of the fully active, mature protein in some areas versus others. Another open question is whether venous and future lymphatic ECs are always fully competent to respond to Vegfc, or whether at certain stages, the response of (future) LECs to Vegfc is modulated, as has been reported for other aspects of the embryonic vascular system (Hogan et al. 2009b).

12.5 The Role of Ccbe1 in Venous Sprouting and Lymphangiogenesis

In a screen for lymphatic mutants, Hogan et al. (2009a) identified a zebrafish mutant that lacked venous sprouts and consequently also all aspects of lymphatic vessel development (Fig. 12.2). Later, other alleles were discovered (Bos et al. 2011) that allowed the formation of some venous sprouts, but still exhibited no signs of lymphatic formation. The causative gene was found to encode Ccbe1, a secreted protein with a calcium-binding EGF domain in the N-terminal part of the protein and a collagen-repeat domain in the C-terminal part. The highly dynamic expression pattern of the gene supports an essential role for *ccbe1* in lymphatic formation: transcript distribution includes (1) the dorsal aspect of the PCV at a time when the first venous sprouts are about to emerge, (2) the migration route of future PL cells in the trunk, and (3) the horizontal myoseptum region where PLs accumulate before they migrate dorsally or ventrally (Hogan et al. 2009a). *Ccbe1* is not expressed in endothelial cells (with very few exceptions), but still appears to influence the migration of LECs in a very direct manner: transcripts are found a few hours before venous sprouts and PLs show migratory behavior in those areas where *ccbe1* is expressed. What controls transcription of *ccbe1* in this tightly controlled manner is not understood, but it has been recently reported that E2f7 and E2f8 influence expression of both *ccbe1* and *flt4* (Weijts et al. 2013).

Ccbe1 function is essential for formation of the lymphatics and this essential requirement is conserved in mammals: *Ccbe1* mutant mice also lack lymphatics (Bos et al. 2011) and even heterozygotes exhibit a phenotype (Hägerling et al. 2013). Furthermore, the human CCBE1 gene is mutated in Hennekam syndrome patients (Alders et al. 2009), a condition that is characterized by lymphedema formation and, among others, cardiovascular anomalies and variable mental retardation (see below and Chap. 14).

Therefore, the situation for CCBE1 is similar to VEGFC and VEGFR-3 in that not only human patients have been described but also that there are mutants in zebrafish and mice that can be studied. Particularly for CCBE1, a lot needs to be learned: it is unclear at present how the protein exerts its effect and how LECs sense the presence of the protein in the environment. Neither has a receptor protein been identified, nor are there other molecular indications described to explain the essential requirement for CCBE1 during lymphatic development. Hägerling et al. (2013) first reported that mice double heterozygous for *Vegf-c* and *Ccbe1* show an exacerbated phenotype compared to single heterozygous situations; since in both zebrafish and mice the respective mutant phenotypes for *Vegf-c* and *Ccbe1* are very similar, this might point towards a common pathway for both genes.

12.6 Zebrafish and Human Disease

There are a number of well-characterized hereditary human syndromes that are mirrored, at least to some extent, by mutant scenarios in the zebrafish.

Milroy disease, characterized by dysfunctioning cutaneous lymphatic vessels, can be caused by missense mutations in the kinase domain of VEGFR-3 (Karkkainen et al. 2000; Mellor et al. 2010) or by heterozygotic loss of VEGFC (Gordon et al. 2013). In the case of VEGFC, so far only one family has been identified and it is an interesting question to consider why it has taken so long to identify human VEGFC patients: In zebrafish forward genetic screens, *vegfc* alleles are frequently found and the loss of a single allele already results in a phenotype (Karpanen and Schulte-Merker, unpublished) as has been described in mice before (Karkkainen et al. 2004; Hägerling et al. 2013), indicating that vertebrate embryos are extremely sensitive to VEGF-C doses. Studies in zebrafish, expressing wild type and the identified mutant c.571_572insTT versions of the human VEGFC proteins in vivo at ectopic locations, have demonstrated that the human c.571_572insTT variant is not acting as a dominant-negative form of VEGFC protein and that human patients, in all likelihood, represent a haploinsufficient situation (Gordon et al. 2013). It is well possible that most human VEGFC/+ situations are early lethal and go unnoticed.

In the case of *ccbe1*, the isolation of the zebrafish mutant through a forward genetic approach (Hogan et al. 2009a) has helped to pinpoint the human orthologue CCBE1 as the gene that is affected in about 40 % of human Hennekam patients (Alders et al. 2009). Particularly since there is the distinct possibility that human patients do not represent full loss-of-function situations, the experimental possibilities and the insight provided through studies in fish and mice have been instrumental to shed light on the function of CCBE1 during embryonic lymphangiogenesis. As other genes must exist that are affected in those Hennekam syndrome patients, which appear to have normal CCBE1 function, one might expect that analysis of these genes in zebrafish and mice will similarly help to understand the human pathology.

12.7 Similarities and Discrepancies

Within the last few years, considerable progress has been made in our understanding of lymphatic development but also in the role of lymphatics in disease (see Chap. 14). By default, this is certainly true not only for lymphangiogenesis in zebrafish with its very short history as an experimental system but also for the murine system where the very first steps of lymphangiogenic sprouting have been reevaluated very recently (Yang et al. 2012; Hägerling et al. 2013). Hence, it is possible now to take a closer look at fish and mice and to ask how similar the development of the lymphatic system in those two key model species really is.

From a geneticist's point of view, there are dominant reasons which support the notion of evolutionary conservation: for VEGFC, FLT4, and CCBE1 there are mutant scenarios in fish, mice, and man (see above). In all cases, there are effects in the lymphatic system, strongly underpinning a common genetic mechanism for the lymphatics to develop. Certainly, there are notable differences: the effect of a heterozygous VEGFC situation in human patients is rather mild (Gordon et al. 2013), while in both mice (Hägerling et al. 2013) and fish (Kärpanen and Schulte-Merker, unpublished) heterozygous situations elicit a more noticeable phenotype. Also, zebrafish and murine Ccbe1 loss-of-function situations effectively lack all lymphatics, while human Hennekam syndrome patients retain some lymphatic structures (Alders et al. 2009). However, there are certain aspects which we probably do not fully understand yet (e.g., it is not clear whether Hennekam syndrome patients represent full loss-of-function situations), and even though there might be differences in details, the overall genetic conservation of VEGFC, FLT4, and CCBE1 function cannot be questioned.

How about some of the other key genes in murine lymphatic development? In mice, Sox18 acts upstream of Prox1 to specify a subset of endothelial cells within the cardinal vein to take on a lymphatic fate. While the role of Sox18 in this process appears to be somewhat dependent on the genetic background, Prox1 is undoubtedly essential for this process (see Chap. 2 and references therein). In zebrafish, the exact function of the *sox 18* and *prox* genes is still an open question: there are reports that have described expression of *prox1a* in lymphatic structures (Yaniv et al. 2006), but knockdown studies using morpholinos result in severely misshapen embryos, complicating an assessment of a requirement for *prox1a*. Similarly, *prox1b* in fish has been claimed to be essential for lymphatic development based on morpholino studies, but subsequent analysis of *prox1b* mutants did not support this notion (Tao et al. 2011). Hence, it will be necessary to generate a *prox1a* zebrafish mutant, and possibly double mutants with *prox1b*, to settle this issue conclusively.

Similarly, a number of studies have examined the role of Sox18 and Coup-TFII in zebrafish. Initially, *sox18* and *sox7* were reported to control arteriovenous specification in a redundant manner, with morpholino knockdown phenotypes showing that both genes regulate the expression of key factors such as *efnb2*, *ephB4* and *flt4* (Cermenati et al. 2008; Herpers et al. 2008; Pendeville et al. 2008). By design, those studies might not necessarily have used morpholino amounts that lead to a complete loss of the respective gene functions. Recently, however, an additional study (Cermenati et al. 2013) reports that morpholino-based knockdown of *sox18* results in lymphatic defects, further supported by results that make use of a dominant-negative version of mouse Sox18-RoA within the zebrafish vasculature: These results are intriguing and seem to be in line with studies that show a similar role for zebrafish Coup-TFII in lymphatic development (Aranguren et al. 2011), but a completely conclusive picture will have to await the generation of stable mutant lines in those genes. Indeed, a zebrafish mutant in *sox18* appears to have no early lymphatic phenotype (Hermkens and Schulte-Merker, unpublished). Of note, there is a requirement in the mouse embryo to specify lymphatic endothelial cells within the existing epithelial sheet of the cardinal vein—in other words, there needs to be a system in place to change the identity of some cells from venous to lymphatic. In fish, venous and lymphatic sprouts appear simultaneously from the posterior cardinal vein (Hogan et al. 2009a), and therefore there might not be a need for the CoupTF/Sox18/Prox1 system to specify a different fate at the level of the cardinal vein.

As discussed above, there is compelling genetic evidence that there are more similarities than differences between the teleost and the murine embryonic lymphatic system. How about the behavior of lymphatic endothelial cells? Here it is interesting to note that the cell mechanics that drive lymphangiogenesis in fish and mice are somewhat different: in zebrafish, sprouts from the PCV emerge at one point in time, and half of these sprouts remodel arteries into veins, while the other half migrates to the horizontal myoseptum to establish a population of individual PL cells. In mice, the formation of lymphatic endothelial cells is chronologically distinct from vein formation. Furthermore, recent studies (Hägerling et al. 2013) making use of high-resolution imaging with an ultramicroscope indicate that future murine LECs emerge from the cardinal vein as a network of cells that are connected to each other. Further studies are required to better understand how these apparent differences between zebrafish and murine lymphangiogenesis on the cellular level can be explained against the background of a conserved genetic program. It may well be that in the case of CCBE1/VEGF-C/VEGFR-3, the common denominator in all vertebrates is to initiate sprouting of future LECs from embryonic veins. All subsequent steps of cellular movements and behavior during embryogenesis have diverged then possibly somewhat during the course of evolution and a better description of the dependence of migratory LECs on extracellular matrix components and cell-cell adhesion molecules is required to allow progress in this fascinating aspect of embryonic lymphangiogenesis.

12.8 Concluding Remarks

Given the brief history of studying lymphatics in zebrafish, we have obtained a remarkably detailed understanding of the cellular behavior of venous and lymphatic endothelial cells during early embryogenesis. Important questions remain to be answered, foremost how venous and lymphatic sprouts are specified during early stages of development. Also, studies have been restricted mainly to embryonic stages, and we only have a cursory picture of adult lymphatic function.

The main advantages of zebrafish, namely, forward genetic screens and in vivo imaging, have been used to full advantage and will undoubtedly continue to provide mutant phenotypes and novel gene functions, as well as detailed insights into the behavior of LECs in vivo. The apparent genetic conservation between teleosts and mammals should allow using zebrafish and mice in a complementary manner also in the future, allowing further progress in what are very exciting times in lymphatics research.

Acknowledgements We would like to thank all lab members, past and present, who have contributed directly, or in discussions, to the work related to lymphatic development. We would like to apologize to authors whose work we could not discuss in detail due to space limitations. Andreas van Impel was supported by a Marie Curie IEF fellowship; Stefan Schulte-Merker is supported by the KNAW.

References

Alders, M., Hogan, B. M., Gjini, E., Salehi, F., Al-Gazali, L., Hennekam, E. A., et al. (2009). Mutations in *CCBE1* cause generalized lymph vessel dysplasia in humans. *Nature Genetics, 41* (12), 1272–1274.

Aranguren, X. L., Beerens, M., Vandevelde, W., Dewerchin, M., Carmeliet, P., & Luttun, A. (2011). Transcription factor COUP-TFII is indispensable for venous and lymphatic development in zebrafish and Xenopus laevis. *Biochemical and Biophysical Research Communications, 410*(1), 121–126.

Bos, F. L., Caunt, M., Peterson-Maduro, J., Planas-Paz, L., Kowalski, J., Karpanen, T., et al. (2011). CCBE1 is essential for mammalian lymphatic vascular development and enhances the lymphangiogenic effect of vascular endothelial growth factor-C in vivo. *Circulation Research, 109*(5), 486–491.

Bussmann, J., Bakkers, J., & Schulte-Merker, S. (2007). Early endocardial morphogenesis requires Scl/Tal1. *PLoS Genetics, 3*(8), e140.

Bussmann, J., Bos, F., Urasaki, A., Kawakami, K., Duckers, H. J., & Schulte-Merker, S. (2010). Arteries provide essential guidance cues for lymphatic endothelial cells in the zebrafish trunk. *Development, 137*, 253–257.

Cermenati, S., Moleri, S., Cimbro, S., Corti, P., Del Giacco, L., Amodeo, R., et al. (2008). Sox18 and Sox7 play redundant roles in vascular development. *Blood, 111*, 2657–2666.

Cermenati, S., Moleri, S., Neyt, C., Bresciani, E., Carra, S., Grassini, D. R., et al. (2013). Sox18 genetically interacts with VegfC to regulate lymphangiogenesis in zebrafish. *Arteriosclerosis, Thrombosis, and Vascular Biology, 33*(6), 1238–1247.

Cha, Y. R., Fujita, M., Butler, M., Isogai, S., Kochhan, E., Siekmann, A. F., et al. (2012). Chemokine signaling directs trunk lymphatic network formation along the preexisting blood vasculature. *Developmental Cell, 22*(4), 824–836.

Covassin, L. D., Villefranc, J. A., Kacergis, M. C., Weinstein, B. M., & Lawson, N. D. (2006). Distinct genetic interactions between multiple Vegf receptors are required for development of different blood vessel types in zebrafish. *Proceedings of the National Academy of Sciences of the United States of America, 103*, 6554–6559.

Dumont, D. J., Jussila, L., Taipale, J., Lymboussaki, A., Mustonen, T., Pajusola, K., et al. (1998). Cardiovascular failure in mouse embryos deficient in VEGF receptor-3. *Science, 282*(5390), 946–949.

Geudens, I., Herpers, R., Hermans, K., Segura, I., Ruiz de Almodovar, C., Bussmann, J., et al. (2010). Role of Dll4/Notch in the formation and wiring of the lymphatic network in zebrafish. *Arteriosclerosis, Thrombosis, and Vascular Biology, 30*(9), 1695–1702.

Gordon, K., Schulte, D., Brice, G., Simpson, M. A., Roukens, M. G., van Impel, A. W., et al. (2013). A mutation in VEGFC, a ligand for VEGFR3, is associated with autosomal-dominant Milroy-like primary lymphedema. *Circulation Research, 112*(6), 956–960.

Hägerling, R., Pollmann, C., Andreas, M., Schmidt, C., Nurmi, H., Adams, R. H., et al. (2013). A novel multi-step mechanism for initial lymphangiogenesis in mouse embryos based on ultramicroscopy. *EMBO Journal, 32*(5), 629–644.

Haiko, P., Makinen, T., Keskitalo, S., Taipale, J., Karkkainen, M. J., Baldwin, M. E., et al. (2008). Deletion of vascular endothelial growth factor C (VEGF-C) and VEGF-D is not equivalent to VEGF receptor 3 deletion in mouse embryos. *Molecular and Cellular Biology, 28*(15), 4843–4850.

Herpers, R., van de Kamp, E., Duckers, H. J., & Schulte-Merker, S. (2008). Redundant roles for sox7 and sox18 in arteriovenous specification in zebrafish. *Circulation Research, 102*, 12–15.

Hewson, W., & Hunter, W. (1769). An account of the lymphatic system in fish. *Philosophical Transactions (1683–1775), 59*, 204–215.

Hogan, B. M., Bos, F., Bussmann, J., Witte, M., Chi, N., Duckers, H., et al. (2009a). Ccbe1 is required for embryonic lymphangiogenesis and venous sprouting. *Nature Genetics, 41*, 396–398.

Hogan, B. M., Herpers, R., Witte, M., Heloterä, H., Alitalo, K., Duckers, H. J., et al. (2009b). Vegfc/Flt4 signalling is suppressed by Dll4 in developing zebrafish intersegmental arteries. *Development, 136*, 4001–4009.

Jeltsch, M., Kaipainen, A., Joukov, V., Meng, X., Lakso, M., Rauvala, H., et al. (1997). Hyperplasia of lymphatic vessels in VEGF-C transgenic mice. *Science, 276*(5317), 1423–1425.

Jensen Dahl Ejby, L., Cao, R., Hedlund, E. M., Söll, I., Lundberg, J. O., Hauptmann, G., et al. (2009). Nitric oxide permits hypoxia-induced lymphatic perfusion by controlling arterial-lymphatic conduits in zebrafish and glass catfish. *Proceedings of the National Academy of Sciences of the United States of America, 106*(43), 18408.

Joukov, V., Sorsa, T., Kumar, V., Jeltsch, M., Claesson-Welsh, L., Cao, Y., et al. (1997). Proteolytic processing regulates receptor specificity and activity of VEGF-C. *EMBO Journal, 16*(13), 3898–3911.

Kampmeier, O. F. (1925). The development of the trunk and tail lymphatics and posterior lymph hearts in anuran embryos. *Journal of Morphology, 41*, 1.

Karkkainen, M. J., Haiko, P., Sainio, K., Partanen, J., Taipale, J., Petrova, T. V., et al. (2004). Vascular endothelial growth factor C is required for sprouting of the first lymphatic vessels from embryonic veins. *Nature Immunology, 5*(1), 74–80.

Karkkainen, M. J., Ferrell, R. E., Lawrence, E. C., Kimak, M. A., Levinson, K. L., McTigue, M. A., et al. (2000). Missense mutations interfere with VEGFR-3 signalling in primary lymphoedema. *Nature Genetics, 25*(2), 153–159.

Karpanen, T., Heckman, C. A., Keskitalo, S., Jeltsch, M., Ollila, H., Neufeld, G., et al. (2006). Functional interaction of VEGF-C and VEGF-D with neuropilin receptors. *FASEB Journal, 20*, 1462–1472.

Küchler, A. M., Gjini, E., Peterson-Maduro, J., Cancilla, B., Wolburg, H., & Schulte-Merker, S. (2006). Development of the zebrafish lymphatic system requires VEGFC signaling. *Current Biology, 16*, 1244–1248.

Mellor, R. H., Hubert, C. E., Stanton, A. W., Tate, N., Akhras, V., Smith, A., et al. (2010). Lymphatic dysfunction, not aplasia, underlies Milroy disease. *Microcirculation, 17*(4), 281–296.

Ny, A., Koch, M., Schneider, M., Neven, E., Tong, R. T., Maity, S., et al. (2005). A genetic Xenopus laevis tadpole model to study lymphangiogenesis. *Nature Medicine, 11*(9), 998–1004.

Okuda, K. S., Astin, J. W., Misa, J. P., Flores, M. V., Crosier, K. E., & Crosier, P. S. (2012). lyve1 expression reveals novel lymphatic vessels and new mechanisms for lymphatic vessel development in zebrafish. *Development, 139*(13), 2381–2391.

Pendeville, H., Winandy, M., Manfroid, I., Nivelles, O., Motte, P., Pasque, V., et al. (2008). Zebrafish Sox7 and Sox18 function together to control arterial-venous identity. *Developmental Biology, 317*, 405–416.

Stacker, S. A., Stenvers, K., Caesar, C., Vitali, A., Domagala, T., Nice, E., et al. (1999). Biosynthesis of vascular endothelial growth factor-D involves proteolytic processing which generates non-covalent homodimers. *Journal of Biological Chemistry, 274*(45), 32127–32136.

Steffensen, J. F., Lomholt, J. P., & Vogel, W. O. P. (1986). In vivo observations on a specialized microvasculature. *Acta Zoologica, 67*(4), 193–200.

Tammela, T., Zarkada, G., Wallgard, E., Murtomaki, A., Suchting, S., Wirzenius, M., et al. (2008). Blocking VEGFR-3 suppresses angiogenic sprouting and vascular network formation. *Nature, 454*(7204), 656–660.

Tao, S., Witte, M., Bryson-Richardson, R. J., Currie, P. D., Hogan, B. M., & Schulte-Merker, S. (2011). Zebrafish *prox1b* mutants develop a lymphatic vasculature, and *prox1b* does not specifically mark lymphatic endothelial cells. *PLoS One, 6*(12), e28934.

Valasek, P., Macharia, R., Neuhuber, W. L., Wilting, J., Becker, D. L., & Patel, K. (2007). Lymph heart in chick–somitic origin, development and embryonic oedema. *Development, 134*(24), 4427–4436.

Villefranc, J. A., Nicoli, S., Bentley, K., Jeltsch, M., Zarkada, G., Moore, J. C., et al. (2013). A truncation allele in vascular endothelial growth factor C reveals distinct modes of signaling during lymphatic and vascular development. *Development, 140*(7), 1497–1506.

Vogel, W. O. P., & Claviez, M. (1981). Vascular specialization in fish, but no evidence for lymphatics. *Zeitschrift für Naturforschung, 36c*, 490–492.

Weijts, B. G. M. W., van Impel, A., Schulte-Merker, S., & de Bruin, A. (2013). Atypical E2fs control lymphangiogenesis through transcriptional regulation of Ccbe1 and Flt4. *PLoS One, 8*(9), e73693.

Yang, Y., Garcia-Verdugo, J. M., Soriano-Navarro, M., Srinivasan, R. S., Scallan, J. P., Singh, M. K., et al. (2012). Lymphatic endothelial progenitors bud from the cardinal vein and intersomitic vessels in mammalian embryos. *Blood, 120*, 2340–2348.

Yaniv, K., Isogai, S., Castranova, D., Dye, L., Hitomi, J., & Weinstein, B. M. (2006). Live imaging of lymphatic development in the zebrafish. *Nature Medicine, 12*(6), 711–716.

Zhang, L., Zhou, F., Han, W., Shen, B., Luo, J., Shibuya, M., et al. (2010). VEGFR-3 ligand-binding and kinase activity are required for lymphangiogenesis but not for angiogenesis. *Cell Research, 20*(12), 1319–1331.

Chapter 13
Visualization of Lymphatic Vessel Development, Growth, and Function

Cathrin Pollmann, René Hägerling, and Friedemann Kiefer

Abstract Despite their important physiological and pathophysiological functions, lymphatic endothelial cells and lymphatic vessels remain less well studied compared to the blood vascular system. Lymphatic endothelium differentiates from venous blood vascular endothelium after initial arteriovenous differentiation. Only recently by the use of light sheet microscopy, the precise mechanism of separation of the first lymphatic endothelial progenitors from the cardinal vein has been described as delamination followed by mesenchymal cell migration of lymphatic endothelial cells. Dorsolaterally of the embryonic cardinal vein, lymphatic endothelial cells reaggregate to form the first lumenized lymphatic vessels, the dorsal peripheral longitudinal vessel and the more ventrally positioned primordial thoracic duct. Despite this progress in our understanding of the first lymph vessel formation, intravital observation of lymphatic vessel behavior in the intact organism, during development and in the adult, is prerequisite to a precise understanding of this tissue. Transgenic models and two-photon microscopy, in combination with optical windows, have made live intravital imaging possible: however, new imaging modalities and novel approaches promise gentler, more physiological, and longer intravital imaging of lymphatic vessels.

13.1 Introduction

Lymphatic vessels, the second branch of the mammalian vascular system, fulfill indispensible functions for tissue fluid homeostasis, trafficking of dendritic cells and macrophages as well as in dietary lipid resorption (Tammela and Alitalo 2010). Recently, the lymphatic vasculature has also been suggested to contribute to pathological processes, including tumor metastasis, inflammatory diseases, cardiac

C. Pollmann • R. Hägerling • F. Kiefer (✉)
Mammalian Cell Signaling Laboratory, MPI for Molecular Biomedicine,
Röntgenstrasse 20, 48149 Münster, Germany
e-mail: fkiefer@gwdg.de

F. Kiefer and S. Schulte-Merker (eds.), *Developmental Aspects of the Lymphatic Vascular System*, Advances in Anatomy, Embryology and Cell Biology 214,
DOI 10.1007/978-3-7091-1646-3_13, © Springer-Verlag Wien 2014

hypertension, and obesity (Alitalo 2011). Despite these important physiological and pathophysiological functions, the lymphatic vessel system has traditionally been considered of "lesser significance" compared to blood vessels. Hence, elucidation of the molecular mechanisms controlling development, maturation, and function of lymphatic vessels for a long time lagged behind the research into blood vessel physiology and function.

A trivial but nevertheless fundamental difference between blood and lymphatic vessels might underlie this lower interest, which is their relative invisibility. Lymphatic capillaries are only surrounded by a thin discontinuous basement membrane, covered by no or few mural cells and in the resting, largely collapsed state carry little colorless fluid (Schulte-Merker et al. (2011); see also Chaps. 3, 4, and 5 this volume). During medical intervention, the need to treat bleeding from injured blood vessels is as obvious as is the necessity to restore blood flow through arteries and veins during surgery, replantation, or transplantation. In contrast, transection of lymphatic vessels has little immediate consequences, and therefore, despite the risk of potentially debilitating long-term consequences, reconnection surgery of the difficult to localize lymphatic vessels is much less frequently performed, albeit this is an established procedure by now (Baumeister and Frick 2003).

Although insight into and understanding of the lymphatic vasculature has made great advances over the last decade, there is a tremendous necessity to visualize lymphatic vessel structure and function in such diverse areas as basic research and applied biomedical investigation but also in the clinical praxis and surgical treatment. Effective and ideally minimally invasive visualization of lymphatic vessels would enormously foster an in-depth understanding of fundamental questions in developmental biology but also aid the development of more efficient and reliable treatment regimes for lymph vessel pathologies.

This review mainly focuses on the visualization of lymphatic endothelium in the model organism mouse, as the observation of lymphatic development in zebrafish has been described in the preceding chapter (see Chap. 12). We will describe recent approaches to analyze embryonic development using wholemount stained, fixed embryos and fetuses at various gestational stages. Here the use of optical sectioning allows the comprehensive spatial interrogation of the entire developing vasculature in unperturbed embryos. In addition, the availability of novel transgenic model systems allows in combination with optical windows the direct and chronic observations of lymphatic vessels under physiological but also pathological conditions. Finally, we will briefly touch upon recent developments, which aim to visualize lymphatic vessels using novel imaging modalities also during diagnosis and therapeutic interventions.

13.2 The First Steps of Lymphatic Vascular Differentiation

In the mouse, development of the lymphatic vascular system starts following arteriovenous differentiation with the specification of lymphatic progenitors in the dorsal half of the common cardinal vein (CCV) (Oliver 2004). The first sign of inequality among venous endothelial cells is expression of Lyve1 around day 9.5 of embryonic development (E9.5) (Fig. 13.1a). Expression of the transcription factor Sox18 is initiated shortly after in these lymphatic competent cells and is the first detectable step in a transcriptional program, which initiates and ultimately establishes lymphatic endothelial cell (LEC) identity (Francois et al. 2008). Sox18, in concert with the nonpolarly expressed venous transcription factor Coup-TFII, directly activates the homeobox transcription factor Prox1 (Srinivasan et al. 2010). Coup-TFII, Sox18, and Prox1 are indispensable for lymphatic fate specification; however, Prox1 alone appears sufficient to drive lymphatic specification and differentiation in vitro (Petrova et al. 2002; Hong et al. 2002). Therefore Prox1 acts as a master regulator of lymphatic differentiation.

Around E10–E10.5, Prox1-expressing lymphatic endothelial progenitors leave the cardinal vein (CV) and establish dorsolaterally the first lumenized lymphatic vascular structures, which were historically referred to as jugular lymph sacs (Sabin 1902; Oliver and Detmar 2002) (Fig. 13.1a). Further lymphatic development happens through sprouting lymphangiogenesis from the first lumenized lymphatic structures. Remarkably, the blood and lymphatic vasculature develop into two completely separate systems except two exclusive contact sites, which form at the junction of the jugular and subclavian veins and are maintained throughout adulthood (Tammela and Alitalo 2010). Anywhere else in the body, despite close proximity, blood and lymphatic vessels possess no direct contact sites.

Over the last years, the exact mechanism how the first lymphatic progenitors separate from the CV has been a matter of discussion. Initial reports suggested that Prox1+ LECs migrate away from the cardinal vein to reorganize into the jugular lymph sacs (Fig. 13.1b track 1) (Wigle and Oliver 1999). An alternative model was sparked by the realization that genetic inactivation of hematopoietic signaling molecules, including the cytoplasmic non-receptor tyrosine kinase Syk and the adaptor protein SLP-76, resulted in the formation of multiple aberrant connections between blood and lymphatic vessels, causing edema and blood-filled lymphatics (Abtahian et al. 2003). The hypothesis that a cell autonomous defect, resulting in aberrant migration of early LEC progenitors, was responsible for the blood-lymphatic shunt formation (Sebzda et al. 2006) was disproved by genetic lineage tracing experiments, which undoubtedly excluded Syk expression in mature or developing endothelial cells (Bohmer et al. 2010). Therefore, the defect in Syk-deficient mice was non-cell autonomous and definitively caused by hematopoietic cells (Bohmer et al. 2010). Shortly after, platelets were identified as the major responsible hematopoietic compartment and it became clear that Syk and

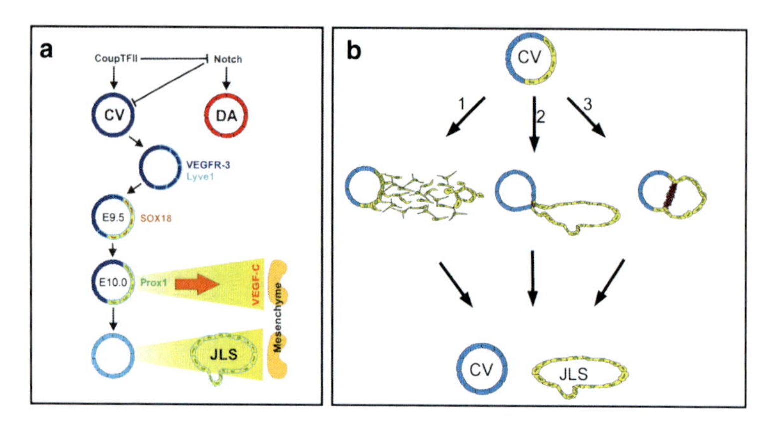

Fig. 13.1 Schematic depiction of lymphatic developmental mechanisms. (**a**) Developmental lymphangiogenesis. After arteriovenous specification VEGFR-3 (*blue*) is highly expressed specifically by venous endothelium. A subpopulation of endothelial cells on the dorsal site of the cardinal vein (CV) starts to express the lymphatic vessel hyaluronan receptor-1 (Lyve1, *light blue*). At E9.5 expression of the transcription factor Sox18 (*red*) is initiated in these lymphatic competent cells, which induces the expression of Prox1 (*green*), the master regulator of lymphatic identity. Concomitantly, VEGFR-3 is downregulated on blood endothelial cells, but remains high on Prox1+ cells and renders them responsive to VEGF-C produced by cells of the lateral mesenchyme. Dorsolaterally of the CV, emigrated LECs organize into lumenized structures, known as the jugular lymph sacs (JLS). (**b**) Conceivable concepts of lymphatic separation. Three concepts for lymphatic separation from the venous endothelium have in the recent past been discussed in the literature. (*1*) Lymphatic endothelial cells leave the CV as single cells that migrate away and reorganize into lumenized structures in the periphery. (*2*) The involvement of platelets in the separation of blood and lymph vessels sparked a new concept, which proposed LECs to emerge from the CV by sprouting. Platelet coagulation at the base of emerging sprouts was hypothesized to trigger the subsequent abscission of newly formed lymphatic vessels from the CV (Kim and Koh 2010; Uhrin et al. 2010). (*3*) Based on OPT reconstructions entire segmental areas of the CV were suggested to balloon out from the CV and pinch off dorsally. Again, aided by platelets, this morphogenetic process would give rise to lymph sacs in a single step (Francois et al. 2012)

SLP-76 act downstream of the procoagulant platelet receptor CLEC-2 (Finney et al. 2012; Uhrin et al. 2010; Bertozzi et al. 2010; Suzuki-Inoue et al. 2010; Carramolino et al. 2010), which binds to and is activated by its cognate ligand podoplanin on lymphatic endothelial and epithelial cells ((Schacht et al. 2003; Suzuki-Inoue et al. 2006), see also Chap. 8).

The realization that CLEC-2-mediated platelet coagulation is indispensable for the establishment and maintenance of blood and lymph vessel separation led to the hypothesis that the first lymphatic vessels emerge by sprouting or ballooning from the CV and that platelets would catalyze the abscission step (Fig. 13.1b, tracks 2 and 3) (D'Amico and Alitalo 2010; Kim and Koh 2010).

13.3 Three-Dimensional Reconstruction of the Complete Embryonic Vasculature Using Ultramicroscopy

The reliable and unambiguous identification of LECs through immunohistological staining for positively identifying markers like Prox1, podoplanin, Lyve-1, and VEGFR-3 is possible for more than a decade. Yet, discrimination between such distinct mechanisms like migration and sprouting of LECs can be extremely challenging on the basis of histological sections. Sectioning and preparation of artifacts often introduce confounding effects that distort the delicate structures of the embryonic vasculature. Depending on the sectional plane, a clear identification of the vessels in a given histological sample may be difficult, if not impossible. In contrast, complete spatial reconstruction of the developing vascular system of an intact embryo would repeal such issues.

Light sheet or planar illumination-based microscopy (PIM) replaces histological by optical sectioning and is particularly well suited for the spatial interrogation of large, complex organ systems (Mertz 2011). We addressed the mechanism of separation of the first LECs from the CV using a PIM variant termed ultramicroscopy, in which an illumination light sheet is moved stepwise through an optically cleared sample (Becker et al. 2008). Fluorescence signals are recorded orthogonally to the light sheet by an area detector, in our case an EM-CCD camera. Presently, the need for optical clearing of the sample precludes the analysis of living material but also the visualization of genetically encoded, fluorescent proteins. In contrast to confocal microscopes, ultramicroscopy works excellent with low numerical aperture and low-magnification lenses, which offer a long working distance and allow the recording of large sample volumes up to 1 cm^3. Ultramicroscopy delivers a perfectly isotropic point spread function; hence, acquired image stacks are ideally suited for 3D volume rendering and can be rotated around all spatial axes without loss of imaging quality.

13.4 The First Lymphatic Endothelial Progenitors Delaminate from the CV and Migrate Away as Non-lumenized Strings of Mesenchymal Cells

Effective exploitation of wholemount stained, intact embryos using ultramicroscopy essentially relies on the capacity to specifically and selectively mark the different branches of the developing vascular system. Of the different available marker molecules, we found combinatorial staining for Prox1, PECAM-1, and VEGFR-3 or the endothelial sialomucin endomucin (Emcn) most informative. The cell surface molecule PECAM-1 is expressed on all vessel types, however, most strongly on arterial vessels. PECAM-1 can therefore be elegantly used as a surrogate arterial marker. During embryonic development, VEGFR-3 marks

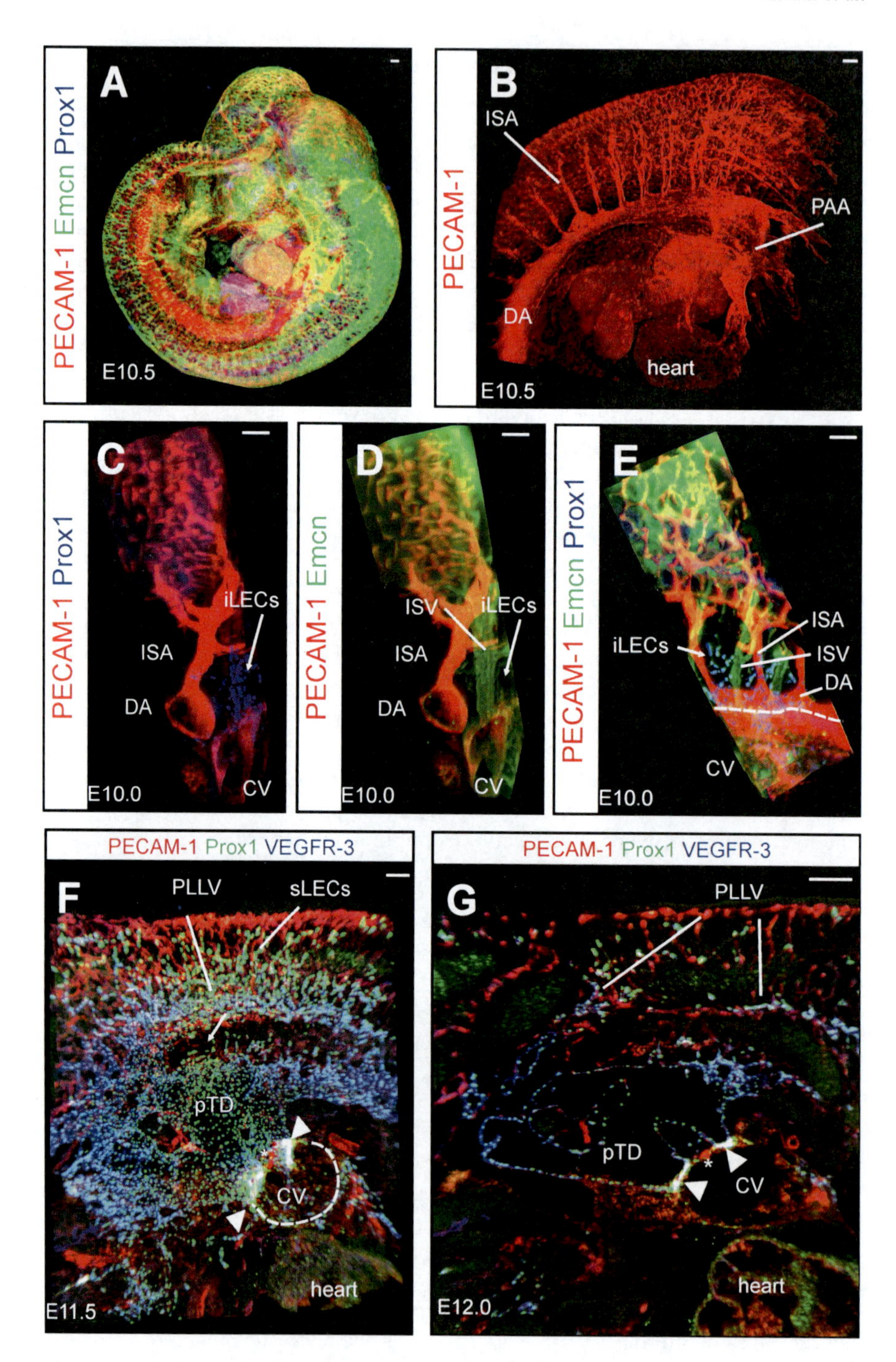

Fig. 13.2 Emergence of the first lymphatic progenitor cells from the cardinal vein. (**a–e**) Wholemount immunostaining of the vasculature in mouse embryos at E10.5 (**a**, **b**) and E10.0 (**c–e**). PECAM-1 preferentially stained arterial, endomucin mostly venous vessels and Prox1

developing veins; however, venous expression is lost following the establishment of the first lymphatic vessels after E11.5. In contrast, Emcn is stably expressed on veins and on the venous side of capillaries throughout development. Similarly, Prox1 identifies LECs at all developmental stages (Fig. 13.2a) (Hägerling et al. 2013).

Three-dimensional (3D) renderings of image stacks acquired by ultramicroscopy provide a complete view of the developing vascular system that offers unprecedented detail and clarity (Fig. 13.2a, b). By ultramicroscopy, data covering the entire embryo are routinely acquired; however, for an in-depth analysis, the data stack is cropped to the area of immediate interest, in our case the jugular-thoracic region for the analysis of the first lymphatic endothelial cells (Fig. 13.2b). In contrast to the analysis of histological sections, in 3D reconstructions the identity of the different vessels can be readily and unequivocally assigned (Fig. 13.2c–e). Ultramicroscopic analysis reveals that immediately after the onset of Prox1 expression, Prox1+ LECs are detected as streams of spindle-shaped, non-lumenized cells located dorsolaterally to the CV (Hägerling et al. 2013). Many of these first LECs, which we termed initial LECs or iLECs, were connected with each other by spot-shaped cell-cell contacts. Although ultramicroscopy does not allow the direct observation of cell migration, their appearance highly suggests that iLECs delaminate from the dorsal roof of the CV and migrate dorsolaterally as streams of mesenchymally shaped cells (Fig. 13.2e). Clearly, analysis of the spatially reconstructed complete developing embryonic vasculature disproved an involvement of sprouting or ballooning mechanisms in the formation of the first lymphatic vessels. Similar results were obtained through the analysis of thick histological sections ((Yang et al. 2012), see also Chap. 2).

Fig. 13.2 (continued) lymphatic endothelial cells (LECs). (**b**) Jugular-thoracic region, in which LECs emerged from the cardinal vein (CV). *DA* dorsal aorta, *ISA* intersegmental artery, *PAAs* pharyngeal arch arteries. Scale bar = 100 μm. (**c–e**) Volume reconstruction of optical sections of embryos wholemount immunostained for the proteins noted on the *left side* of each panel. ((**c**, **d**) transverse sections; (**e**) sagittal section) Prox1+ iLECs (*arrows*) emerge from the dorsal roof of the CV (*dotted line*). ISV, intersegmental vein. Scale bar = 100 μm. (**f**, **g**) Formation of the PLLV and pTD. Paired contact sites between the newly formed pTD and the CV are characterized by highest levels of Prox1 expression (*arrowheads*). Sagittal view of wholemount immunostained embryos. (**f**) The newly formed pTD rapidly consolidated into a massive lumenized structure, cranially connected in a U shape to the PLLV (*left side* (**f**)). (**g**) A transient side branch of the subclavian artery was stereotypically located between the contacts of pTD and CV and is marked by an asterisk. (**g**) Individual plane (optical section) through the contact area of pTD and CV. sLECs, superficial LECs. Scale bars = 100 μm. (Modified from (Hägerling et al. 2013))

13.5 Lymphatic Endothelial Progenitors Condense into Two Large Lumenized Vessels: The Peripheral Longitudinal Lymphatic Vessel and the Primordial Thoracic Duct

While LECs emerge on the dorsal site of the CV, the Prox1 expression domain transiently expands over the entire CCV. Factors, determining which cells leave the cardinal vein, are presently unclear. Compared to surrounding tissue, there is little proliferation within the iLEC population inside and outside the CV. This suggests that the overwhelming majority of iLECs, which rapidly fill the space between the CV and the first lateral side branch of the intersegmental vessels, are derived by emigration from the CV. Within the next 24 h iLECs condense and lumenize, which results in the formation of two distinct lymphatic vessels. The smaller vessel is located dorsally, underneath the ventral edge of the superficial venous plexus (sVP), while the larger vessel is situated more ventrally, close to the CV. We termed the more dorsally located vessel peripheral longitudinal lymphatic vessel (PLLV) (Fig. 13.2f, g). Originating from the PLLV, the superficial lymphatic endothelial cells (sLECs) in the skin form through massive radial sprouting. The more ventrally situated the vessel is the largest lymphatic structure in the midgestation embryo and we termed it primordial thoracic duct (pTD). The pTD forms two contact sites with the CV, which are characterized by a double layer of highly Prox1+ ECs. These contact areas are stereotypically separated by a side branch of the subclavian artery and will subsequently form the lymphovenous valves (Hägerling et al. 2013; Srinivasan and Oliver 2011).

13.6 An Additional Venous Source of Lymphatic Endothelial Progenitors Is Revealed by Loss of Either the Matrix Component CCBE1 or the Growth Factor VEGF-C

Collagen- and calcium-binding EGF domains 1 (CCBE1) is a secreted molecule, which bears resemblance to extracellular matrix components and binds to the matrix protein vitronectin (Bos et al. 2011). CCBE1 mutations cluster in Hennekam syndrome patients, who suffer from a rare multi symptom disease that is associated with lymphedema and lymphangiectasias (Van Balkom et al. 2002). In mice and zebrafish, genetic deletion of CCBE1 causes the loss of embryonic lymphangiogenesis (Bos et al. 2011). Our ultramicroscopic analysis revealed that in CCBE1 mutant mouse embryos, Prox1+ LECs are specified properly in the CV. However, iLECs are completely absent and multiple blind-ended, aberrant sprouts protrude from the CV (Fig. 13.3b, arrowheads). These observations suggest that in the absence of CCBE1, LECs fail to undergo the endothelial to mesenchymal transition (EndMT), which allows them to leave the CV and to migrate dorsally by

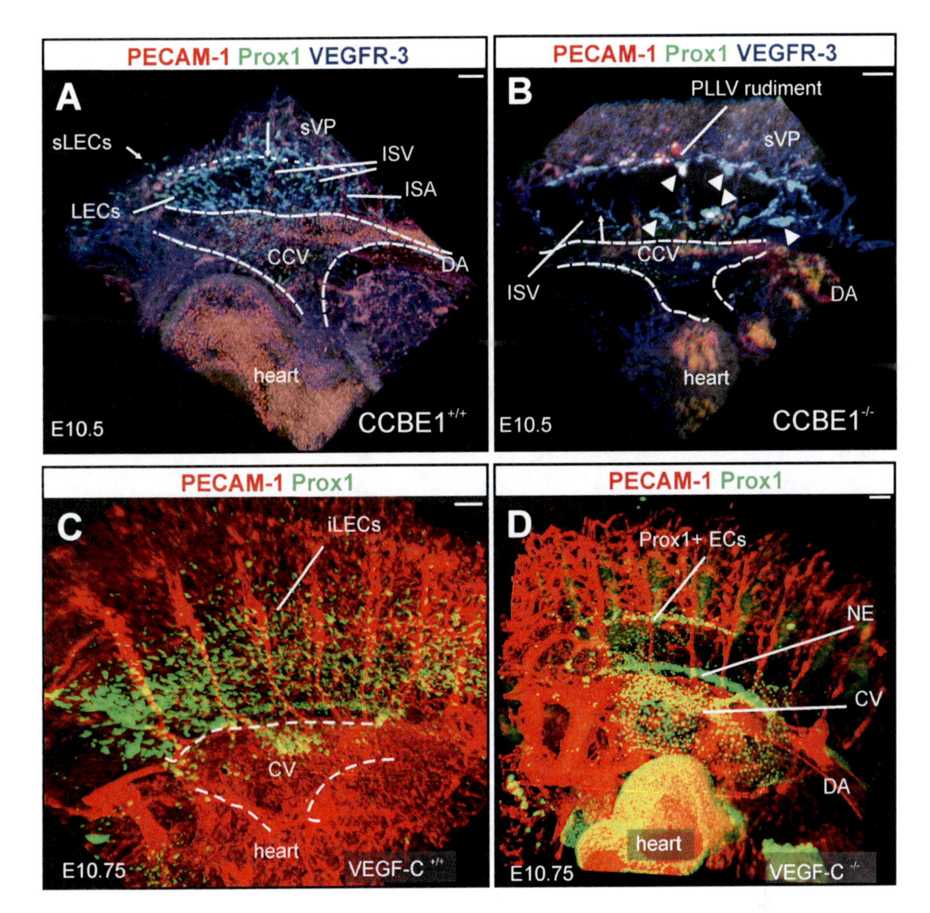

Fig. 13.3 Development of the lymphatic vasculature is impaired in CCBE1 and in VEGF-C-deficient embryos. (**a**, **b**) Sagittal view on 3D reconstructions of wild-type (**a**) and *Ccbe1*$^{-/-}$ (**b**) embryos at E10.5, wholemount immunostained for the indicated proteins. (**b**) CCBE1 deficiency results in a failure of Prox1+ cells to separate from the CV and in rapid loss of nascent lymphatic structures. Abundant Prox1+ cells were detected in the CV and in a rudimentary PLLV, adjacent to the superficial venous plexus (sVP). In contrast to wild-type embryos (**a**), the meshwork of spindle-shaped iLECs between CCV and PLLV was absent. Prox1+ cells outline the CCV and SV, while atypical, large, lumenized sprouts emerged from the CV (*arrowheads*). Scale bars = 100 μm. (**c**, **d**) Also in VEGF-C-deficient embryos, Prox1+ cells accumulate in the CV and ventrally adjacent to the sVP. (Modified from (Hägerling et al. 2013))

mesenchymal cell streaming. Therefore, CCBE1 is an important factor during EndMT. In addition, the analysis of CCBE1 mutant embryos sparked another important insight. Due to the failure of iLECs to delaminate from the CCV, the region between the CV and the ventral edge of the sVP does not contain any LECs. Yet, a rudimentary form of the PLLV can be recognized, which obviously is unlikely to originate from the CV. This observation triggered the assumption that beside the CV there may be at least one additional source of Prox1+ LECs.

Indeed, the analysis of VEGF-C-deficient mice confirmed this hypothesis (Karkkainen et al. 2004). Also in the absence of the promigratory growth factor VEGF-C, Prox1+ LECs are properly specified in the CV. Because of the lack of guidance and migratory cues provided by VEGF-C, LECs fail to delaminate and remain stuck at their place of specification (Fig. 13.3c, d). VEGF-C mutants therefore trap LECs in their venous origin and indeed, we identified the ventral edge of the sVP as a second source of Prox1+ LECs, independent of the CV (Hägerling et al. 2013).

13.7 The Function of Platelets During the Separation of Blood and Lymphatic Vessels Remains Unclear

Our analysis of the first steps of lymphatic vessel formation clearly disproved the hypothesis that platelets partake in or facilitate the separation of sprouting LECs from the CV. Interestingly, iLECs are Pdpn negative and only start to express this protein, after they become incorporated into lumenized lymphatic structures, which certainly protects the embryo against inappropriate coagulation within the blood vasculature (May et al. 2009). However, the function of platelets during the maintenance of blood and lymphatic vessel separation remains enigmatic.

Without question, the Pdpn-CLEC2 axis provides a safety mechanism, which occludes inappropriate connections between previously separate blood and lymphatic vessels or blood vessels and epithelial surfaces and thereby prevents bleeding into the lymphatic vasculature or body cavities. In agreement with this function, we have previously identified the formation of blood-lymphatic shunts in the skin of Syk-deficient fetuses (Bohmer et al. 2010). How often and when these connections form, where exactly they are located, and if they are favored by the relative absence of mural cells and extracellular matrix proteins in the midgestation embryo remains unknown. Interestingly, transplantation of Syk-deficient fetal liver cells into lethally irradiated adult mice indicated that the Pdpn-CLEC2-mediated separation of blood and lymph vessels is active throughout life (Kiefer et al. 1998). Reconstruction of the developing lymphatic system of Syk and SLP-76-defficient embryos during various embryonic and fetal stages using ultramicroscopy should provide important insights into this process.

Recent experiments using "knock-in" mice with altered signaling molecules in myeloid cells and platelets raise the intriguing possibility that platelets exert functions beyond a mere coagulative occlusion of blood-lymphatic shunts. Platelets from mice, in which Syk was replaced by the homologous kinase Zap-70, are completely devoid of CLEC-2-mediated signaling, yet these animals only rarely and at late age suffer from hemorrhages due to blood-lymphatic shunting, suggesting that platelets might influence lymphatic vessel growth, beyond Pdpn-triggered coagulation (Konigsberger et al. 2012).

13.8 Current Limitations of Light Sheet Microscopy

Presently, for opaque model organisms, ultramicroscopy is limited to fixed, immunostained and optically cleared samples. In the mouse, efficient and reliable wholemount immunostaining is only possible until about E12.5. In fetuses or larger specimen, tissue penetration of antibodies becomes the limiting factor, resulting in unreliable staining patterns. A potential solution for this limitation is the use of transgenic model systems, which express genetically encoded fluorescent proteins (FPs) under the control of tissue-specific promoter elements. Unfortunately, the majority of published tissue clearing protocols is either not at all or only to a very limited degree capable of retaining the activity of FPs. Somewhat disappointing, also protocols that were designed to retain the fluorescence of GFP or RFP and their variants rely on very high initial expression, as most of the activity is lost during the clearing procedure (Hama et al. 2011; Becker et al. 2012; Erturk et al. 2012; Kuwajima et al. 2013). Very recently, a promising novel innovative procedure became available under the acronym clarity (Chung et al. 2013), which is now intensely tested by many laboratories.

Despite the possibility to prepare a large range of developmentally closely timed samples, the analysis of fixed specimen is intrinsically inferior to dynamic data derived from living embryos and "sensu stricto" precludes a direct analysis of cell migration. PIM has been extensively used to analyze the development of translucent organisms like zebrafish (Huisken et al. 2004). Recent developments now promise that in the future also in opaque embryos, processes that take place in proximity to the body surface, like the delamination and migration of iLECs from the CV, will become amenable to life PIM. First, new brightly fluorescent transgenic models, in which specifically LECs are labelled by various FPs, have become available (see Sect. 13.10.3). Second, conditions have been described that sustain embryonic development ex utero for extended periods of time (Zeeb et al. 2012) and finally a light sheet microscope specifically dedicated for the intravital use in non-translucent tissue is presently under development.

13.9 Approaches for the Dynamic Visualization of Lymphatic Vessels in Living Organisms

In the healthy adult, the lymphatic vasculature appears to be a stable organ; however, it is subject to constant regulation and maintenance. We have already pointed out in this chapter that the separation of blood and lymphatic vessels is a lifelong active process that relies on CLEC-2-mediated platelet functions. In addition, active lymphangiogenesis and lymphatic vessel remodelling are associated with or may causally contribute to pathologies like edema, inflammation, and tumor metastasis, but in many pathological conditions, the functional involvement of lymphatic vessels is far from clear. In inflammatory and autoimmune processes lymphatic vessels have been considered to fulfill an aggravating function, because as part of the effector arm of the immune response, they ensure efficient transport of antigenic material to the regional lymph nodes (Ingersoll et al. 2011; Kholová et al. 2011; Alitalo 2011; Cursiefen et al. 2004; see also Chap. 10). At the same time, other model systems provided evidence for an ameliorating function of lymphatic vessels by reducing tissue edema and providing a route for efficient drainage of inflammatory myeloid cells (Huggenberger et al. 2011; Huggenberger et al. 2010).

An increasing number of molecular targets that potentially allow the pharmacological modulation of vascular growth and function are presently being explored in various therapeutic settings. Due to many shared regulatory mechanisms, however, the selective modification of either the blood or lymphatic vascular system remains so far an unaccomplished feat (Simons and Eichmann 2013). Indeed, any therapy aiming to specifically target the blood endothelial cells must also be carefully screened for effects on the lymphatic endothelium. However, an in-depth understanding of the function and regulation of lymphatic vessels will not be possible without dynamic, ideally long-term intravital observation of the lymphatic vasculature.

13.9.1 Acute Versus Chronic Intravital Imaging

We distinguish here acute from chronic intravital imaging and will focus on the latter. In the mouse model, acute intravital imaging covers a single imaging session and is largely limited by the time over which anesthesia can be maintained without interfering with vital functions or fundamental biological processes. In contrast, chronic imaging allows the repetitive observation, during temporally limited imaging session, of the same vessel bed over periods of days, weeks, or even months. Only chronic intravital imaging allows the observation of vessel behavior over developmentally or pathophysiologically relevant time frames in a meaningful manner.

13.10 Chronic Intravital Imaging of Lymphatic Vessels

Central to the development of chronic intravital imaging is the establishment of treatment protocols that are not only mild enough to allow survival and full regeneration of the model organism after each imaging session, but they at the same time are also neutral towards the observed organ system. The procedures must not provoke changes in the lymphatic vasculature per se, for instance, through the induction of inflammation, changes in blood flow, or through tissue compression. Three basic parameters, which are tightly interconnected and have to be carefully chosen, define chronic imaging models and the quality of information that can be derived from each model: first, the imaging modality, second, the choice of observation field, and third, the method of contrast generation.

13.10.1 Different Modalities in Lymphatic Vessel Imaging

Presently, optical—in particular fluorescence-based—imaging approaches provide the highest degree of resolution and are simultaneously capable of delivering molecular information on multiple subcellular structures or cell types, however, at the expense of limited penetration depth and mostly small observation fields.

Due to light scattering in living tissue, only confocal microscopy is applicable to intravital imaging of opaque model organisms like the mouse. In particular two-photon laser scanning microscopy (2P-LSM) is widely used, because illumination with femtosecond-pulsed infrared high-energy laser light used in 2P-LSM offers deeper tissue penetration and less phototoxicity compared to fluorescence excitation with visible continuous-wave laser light used in one-photon laser scanning microscopy. The relatively slow acquisition speed of single-beam scanning systems that is usually applied in 2P-LSM is not limiting during the observation of slow processes like vascular behavior or development. For more specialized applications, like the measurement of blood flow velocity, scan speed can be improved through the use of line scanning systems, which illuminate the sample with a line of up to 64 laser light focus points. This however limits detection to the use of area detectors like EM-CCD cameras (Niesner et al. 2008).

Optical projection tomography (OPT) combines bright field or fluorescence images taken at different angles into a single 3D reconstruction. Together with the already-discussed light sheet or planar illumination microscopy (PIM) OPT offers minimal phototoxicity and allows the observation of large sample volumes, like complete embryos and fetuses (Quintana and Sharpe 2011). Presently, both technologies are only suited for optically translucent model organisms; however, attempts to construct a confocal PIM that is applicable to opaque life tissue are under way and the prototype is eagerly awaited (Volker Andresen LaVision Biotech personal communication). Promising, partly still experimental imaging approaches include optical coherence tomography (OCT) (Vakoc et al. 2012) and laser speckle imaging (LSI) (Kalchenko et al. 2012), which are based on the analysis of light propagation and scattering in tissue and are therefore label-free imaging modalities that exploit the contrast of intrinsic tissue properties. Finally, photoacoustic microscopy or tomography (PAM/PAT) is an emerging technology that bears the promise to ultimately deliver optical resolution on a mesoscopic scale. Principle is the spatially resolved detection of ultrasound waves in the MHz range that are generated by the transient thermoelastic expansion of light-absorbing structures, when tissue is illuminated by nanosecond-pulsed laser light (Ntziachristos 2010; Laufer et al. 2012). Because photoacoustic imaging listens to the absorption of light energy, it will be important to develop chemically tailored dye molecules that are particularly efficient in the conversion of light into sound and sufficiently different in the generated acoustic waves to be used in combination, i.e., to allow polyphonic detection. We propose to term such a new class of molecules sonophores.

13.10.2 The Choice of Observation Field

The limited penetration depth of optical microscopy defines narrow constraints for possible observation fields in the intravital imaging of lymphatic vessels. In mice, the dense fur and associated hair follicles limit noninvasive microscopy approaches to three areas, the skin of the ear and tail (e.g., see Proulx et al. 2013; Kwon and Sevick-Muraca 2010), on which hair density is low enough to allow direct access, and the cornea (Yuen et al. 2011; Steven et al. 2011). While limitations arising from the fur in principle can be bypassed by the use of nude mice, only few genetic models, which are mostly generated on the B6 or 129 background, are available as backcross to the nude strain. The dark skin pigment of B6 and 129 mice is also a confining issue for 2P-LSM, where the dark pigments efficiently absorb NIR light and severely limit the access through the skin surface.

In contrast to the intact skin, the cornea offers a particularly well-suited imaging model, as it is inherently transparent and vascular (see Chap. 10 for a detailed review). Ingrowth of lymphatic vessels in response to angiogenic and lymphangiogenic stimuli happens in a predictable and planar fashion from the limbal vessels and is therefore well accessible to 2P-LSM. In particular 2P-LSM but also OCT can be very effectively combined with endoscopy techniques and optical windows. The most extensively used observation system is the dorsal skin fold chamber, which was initially designed for intravital microscopy of the microcirculation under various conditions (Lehr et al. 1993) and offers distinct advantages, as it stably overcomes the skin barrier and allows the controlled and localized application of substances in or close to the field of observation.

13.10.3 Methods of Contrast Generation

Most imaging modalities rely on the availability of substances, which not only introduce contrast into tissue but also provide selectivity to the detection process. These include free contrasting agents, specific high-affinity ligands coupled to a contrast-generating moiety or easily detectable proteins expressed under a specific promoter. In particular, the combination of 2P-LSM and genetically encoded FPs has developed into a workhorse for intravital chronic imaging. Within the last 3 years, four transgenic model systems have become available that express FPs in a lymphendothelial-specific fashion. In one case, a GFP-lacZ fusion protein is expressed under the control of the *Vegfr-3* locus (Martinez-Corral et al. 2012); in the other cases the FPs GFP (Choi et al. 2011), mOrange2 (Hägerling et al. 2011), and tdTomato (Truman et al. 2012) are expressed under the control of Prox1 promoter elements. Efficient use of the red-shifted fluorescent ds-Red derivatives like mOrange2 and tdTomato, which are all optimally excited at 1,100 nm, a wavelength outside the tuning range of pulsed femtosecond IR lasers, has become possible through the availability of optical parametric oscillators (Herz et al. 2010).

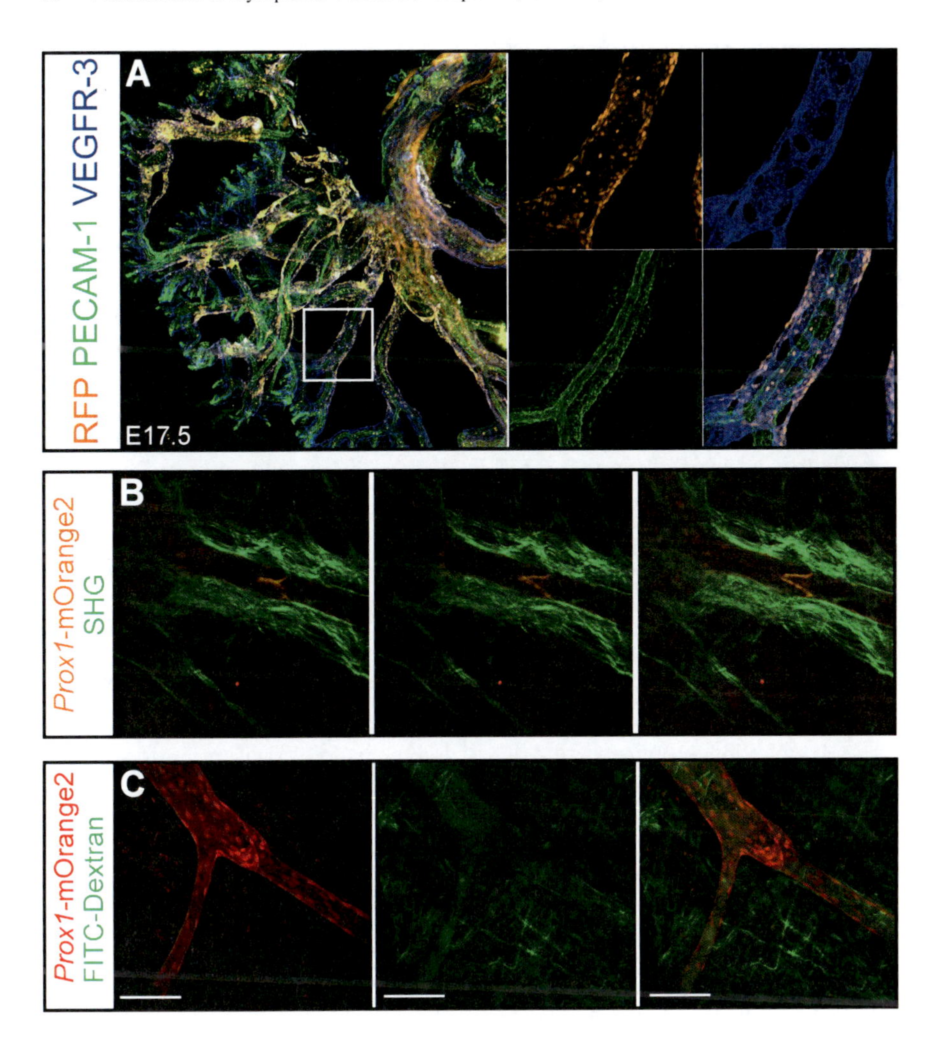

Fig. 13.4 The *prox1*-mOrange2-pA-BAC reporter mouse specifically labels lymphatic endothelium. (**a**) Wholemount immunostaining of embryonic mesentery (E17.5) of the *prox1*-mOrange2-pA-BAC reporter mouse with positively identifying markers. Staining with anti-RFP antibody delineates mOrange2^{+} cells and perfectly colocalizes with PECAM1^{low} and VEGFR-3^{+} structures. (**b**) 2P-LSM intravital microscopy of lymphatic function and valve action. A series of images is presented showing the opening of a lymphatic valve upon lymph flow in the skin of an adult mouse. Lymphatic vessels are marked by mOrange2 expression and the surrounding tissue is visualized through second harmonic signals (SHG) from the extracellular matrix. (**c**) Lymphatic valve architecture and lymph flow after FITC (tracer) injection. 2P images visualize a lymphatic collecting vessel including a saddle-shaped lymphatic valve (*left* panel, *red*) and FITC uptake into a draining lymphatic vessels (*central* panel, *green* content within the vessel) which is embedded in abundant connective tissue (*right-hand* panel, merged image showing FITC-filled lymphatic vessel and ECM of the surrounding tissue)

Hence, the simultaneous detection of three or four genetically encoded FPs in a single multicolor application has become possible. A further distinct advantage of genetically encoded FPs is the regeneration of the labelling system between subsequent imaging sessions.

We have extensively used a (BAC) *prox1*-mOrange2pA reporter mouse, which directly labels LECs with subcellular resolution (Fig. 13.4a), in conjunction with the dorsal skin fold chamber and 2P-LSM to follow lymphangiogenic processes as the development and function of lymphatic valves (Fig. 13.4b, c) (Hägerling et al. 2011). Using this system, we were able to visualize the function of lymphatic valves in the skin of adult mice (Fig. 13.4b, c; see also Chap. 6 of this issue) (Sabine et al. 2012).

The most straightforward form of contrast generation is the simple intradermal injection of a tracer dye, which is subsequently taken up into the lymphatic vasculature and imaged through the skin. While this method delivers functional information on the draining lymphatics, it only provides an outline of the lymphatic vessel volume and relatively low resolution. Depending on the side of injection the lymphatic vessels may fill unevenly, or residual tracer may interfere with or obstruct important areas during imaging sessions. More recently, intradermal application of suitably colored tracers is used in combination with genetically encoded FPs (Fig. 13.4c).

A special case is the injection of the near-infrared dye indocyanine green (ICG) followed by noninvasive detection, which is also directly applicable to humans to assess lymph flow/propulsion and localize lymphatic vessels prior to surgery (Rasmussen et al. 2009; Kwon and Sevick-Muraca 2007; Sevick-Muraca et al. 2008). New improved near-infrared dyes are presently being tested (Proulx et al. 2013).

Finally, the well-established clinical, noninvasive, whole-body imaging modalities positron-emission tomography (PET), single-photon emission computed tomography (SPECT), computer tomography (CT), and magnetic resonance imaging (MRI) can be used to visualize lymphatic vessels. The strengths of these modalities are their virtually unlimited imaging depth and field of view. However, in the mouse they not or only barely resolve small capillaries. Improvement of resolution can be achieved by the use of lymph vessel-specific (e.g., immunoPET (Mumprecht et al. 2010) or bimodal probes (Mounzer et al. 2007) and coregistration of MRI/FRI or PET/CT.

13.11 Implications and Future Directions

An in-depth understanding of the development but also physiology and pathophysiology of lymphatic vessels will not be possible without the dynamic observation of lymph vessels in the living organism. Presently, optical imaging still offers superior resolution and the distinction of multiple molecular targets, but is severely limited in opaque tissue. The analysis of embryonic lymph vessel development should benefit from the availability of improved "ex utero" culture techniques in combination with advanced microscopic modalities that will use numerical methods to deconvolute data obtained from scattering tissue. At the same time, advanced genetic models, more specific intravital probes, novel endoscopic technologies, and the coregistration of different imaging modalities should benefit the analysis of lymphatic vessels in the adult model organism and ultimately in clinical settings.

Acknowledgements Work in the author's laboratory has been supported by the Max Planck Society and the Deutsche Forschungsgemeinschaft SFB 629 and SFB 656. We apologize to many colleagues whose important work could not be cited due to space restrictions. The authors declare no conflicts of interest.

References

Abtahian, F., Guerriero, A., Sebzda, E., Lu, M. M., Zhou, R., Mocsai, A., et al. (2003). Regulation of blood and lymphatic vascular separation by signaling proteins SLP-76 and Syk. *Science, 299*, 247–251.

Alitalo, K. (2011). The lymphatic vasculature in disease. *Nature Medicine, 17*, 1371–1380.

Baumeister, R. G., & Frick, A. (2003). [The microsurgical lymph vessel transplantation]. *Handchirurgie, Mikrochirurgie, Plastische Chirurgie, 35*, 202–209.

Becker, K., Jahrling, N., Kramer, E. R., Schnorrer, F., & Dodt, H. U. (2008). Ultramicroscopy: 3D reconstruction of large microscopical specimens. *Journal of Biophotonics, 1*, 36–42.

Becker, K., Jahrling, N., Saghafi, S., Weiler, R., & Dodt, H. U. (2012). Chemical clearing and dehydration of GFP expressing mouse brains. *PLoS One, 7*, e33916.

Bertozzi, C. C., Schmaier, A. A., Mericko, P., Hess, P. R., Zou, Z., Chen, M., et al. (2010). Platelets regulate lymphatic vascular development through CLEC-2-SLP-76 signaling. *Blood, 116*, 661–670.

Bohmer, R., Neuhaus, B., Buhren, S., Zhang, D., Stehling, M., Bock, B., et al. (2010). Regulation of developmental lymphangiogenesis by Syk(+) leukocytes. *Developmental Cell, 18*, 437–449.

Bos, F. L., Caunt, M., Peterson-Maduro, J., Planas-Paz, L., Kowalski, J., Karpanen, T., et al. (2011). CCBE1 is essential for mammalian lymphatic vascular development and enhances the lymphangiogenic effect of vascular endothelial growth factor-C in vivo. *Circulation Research, 109*, 486–491.

Carramolino, L., Fuentes, J., García-Andrés, C., Azcoitia, V., Riethmacher, D., & Torres, M. (2010). Platelets play an essential role in separating the blood and lymphatic vasculatures during embryonic angiogenesis. *Circulation Research, 106*, 1197–1201.

Choi, I., Chung, H. K., Ramu, S., Lee, H. N., Kim, K. E., Lee, S., et al. (2011). Visualization of lymphatic vessels by Prox1-promoter directed GFP reporter in a bacterial artificial chromosome-based transgenic mouse. *Blood, 117*, 362–365.

Chung, K., Wallace, J., Kim, S. Y., Kalyanasundaram, S., Andalman, A. S., Davidson, T. J., et al. (2013). Structural and molecular interrogation of intact biological systems. *Nature, 497*, 332–337.

Cursiefen, C., Chen, L., Borges, L. P., Jackson, D., Cao, J., Radziejewski, C., et al. (2004). VEGF-A stimulates lymphangiogenesis and hemangiogenesis in inflammatory neovascularization via macrophage recruitment. *Journal of Clinical Investigation, 113*, 1040–1050.

D'Amico, G., & Alitalo, K. (2010). Inside bloody lymphatics. *Blood, 116*, 512–513.

Erturk, A., Mauch, C. P., Hellal, F., Forstner, F., Keck, T., Becker, K., et al. (2012). Three-dimensional imaging of the unsectioned adult spinal cord to assess axon regeneration and glial responses after injury. *Nature Medicine, 18*, 166–171.

Finney, B. A., Schweighoffer, E., Navarro-Núñez, L., Bénézech, C., Barone, F., Hughes, C. E., et al. (2012). CLEC-2 and Syk in the megakaryocytic/platelet lineage are essential for development. *Blood, 119*, 1747–1756.

Francois, M., Caprini, A., Hosking, B., Orsenigo, F., Wilhelm, D., Browne, C., et al. (2008). Sox18 induces development of the lymphatic vasculature in mice. *Nature, 456*, 643–647.

Francois, M., Short, K., Secker, G. A., Combes, A., Schwarz, Q., Davidson, T. L., et al. (2012). Segmental territories along the cardinal veins generate lymph sacs via a ballooning mechanism during embryonic lymphangiogenesis in mice. *Developmental Biology, 364*, 89–98.

Hägerling, R., Pollmann, C., Andreas, M., Schmidt, C., Nurmi, H., Adams, R. H., et al. (2013). A novel multistep mechanism for initial lymphangiogenesis in mouse embryos based on ultramicroscopy. *EMBO Journal, 32*(5), 629–644.

Hägerling, R., Pollmann, C., Kremer, L., Andresen, V., & Kiefer, F. (2011). Intravital two-photon microscopy of lymphatic vessel development and function using a transgenic Prox1 promoter-directed mOrange2 reporter mouse. *Biochemical Society Transactions, 39*, 1674–1681.

Hama, H., Kurokawa, H., Kawano, H., Ando, R., Shimogori, T., Noda, H., et al. (2011). Scale: A chemical approach for fluorescence imaging and reconstruction of transparent mouse brain. *Nature Neuroscience, 14*, 1481–1488.

Herz, J., Siffrin, V., Hauser, A. E., Brandt, A. U., Leuenberger, T., Radbruch, H., et al. (2010). Expanding two-photon intravital microscopy to the infrared by means of optical parametric oscillator. *Biophysical Journal, 98*, 715–723.

Hong, Y. K., Harvey, N., Noh, Y. H., Schacht, V., Hirakawa, S., Detmar, M., et al. (2002). Prox1 is a master control gene in the program specifying lymphatic endothelial cell fate. *Developmental Dynamics, 225*, 351–357.

Huggenberger, R., Siddiqui, S. S., Brander, D., Ullmann, S., Zimmermann, K., Antsiferova, M., et al. (2011). An important role of lymphatic vessel activation in limiting acute inflammation. *Blood, 117*, 4667–4678.

Huggenberger, R., Ullmann, S., Proulx, S. T., Pytowski, B., Alitalo, K., & Detmar, M. (2010). Stimulation of lymphangiogenesis via VEGFR-3 inhibits chronic skin inflammation. *Journal of Experimental Medicine, 207*, 2255–2269.

Huisken, J., Swoger, J., Del Bene, F., Wittbrodt, J., & Stelzer, E. H. K. (2004). Optical sectioning deep inside live embryos by selective plane illumination microscopy. *Science, 305*, 1007–1009.

Ingersoll, M. A., Platt, A. M., Potteaux, S., & Randolph, G. J. (2011). Monocyte trafficking in acute and chronic inflammation. *Trends in Immunology, 32*, 470–477.

Kalchenko, V., Kuznetsov, Y., Meglinski, I., & Harmelin, A. (2012). Label free in vivo laser speckle imaging of blood and lymph vessels. *Journal of Biomedical Optics, 17*, 050502.

Karkkainen, M. J., Haiko, P., Sainio, K., Partanen, J., Taipale, J., Petrova, T. V., et al. (2004). Vascular endothelial growth factor C is required for sprouting of the first lymphatic vessels from embryonic veins. *Nature Immunology, 5*, 74–80.

Kholová, I., Dragneva, G., Cermakova, P., Laidinen, S., Kaskenpää, N., Hazes, T., et al. (2011). Lymphatic vasculature is increased in heart valves, ischaemic and inflamed hearts and in cholesterol-rich and calcified atherosclerotic lesions. *European Journal of Clinical Investigation, 41*, 487–497.

Kiefer, F., Brumell, J., Al-Alawi, N., Latour, S., Cheng, A., Veillette, A., et al. (1998). The Syk protein tyrosine kinase is essential for Fcγ receptor signaling in macrophages and neutrophils. *Molecular and Cellular Biology, 18*, 4209–4220.

Kim, H., & Koh, G. Y. (2010). Platelets take the lead in lymphatic separation. *Circulation Research, 106*, 1184–1186.

Konigsberger, S., Prodohl, J., Stegner, D., Weis, V., Andreas, M., Stehling, M., et al. (2012). Altered BCR signalling quality predisposes to autoimmune disease and a pre-diabetic state. *EMBO Journal, 31*, 3363–3374.

Kuwajima, T., Sitko, A. A., Bhansali, P., Jurgens, C., Guido, W., & Mason, C. (2013). ClearT: A detergent- and solvent-free clearing method for neuronal and non-neuronal tissue. *Development, 140*, 1364–1368.

Kwon, S., & Sevick-Muraca, E. M. (2007). Noninvasive quantitative imaging of lymph function in mice. *Lymphatic Research and Biology, 5*, 219–231.
Kwon, S., & Sevick-Muraca, E. M. (2010). Functional lymphatic imaging in tumor-bearing mice. *Journal of Immunological Methods, 360*, 167–172.
Laufer, J., Norris, F., Cleary, J., Zhang, E., Treeby, B., Cox, B., et al. (2012). In vivo photoacoustic imaging of mouse embryos. *Journal of Biomedical Optics, 17*, 061220.
Lehr, H. A., Leunig, M., Menger, M. D., Nolte, D., & Messmer, K. (1993). Dorsal skinfold chamber technique for intravital microscopy in nude mice. *American Journal of Pathology, 143*, 1055–1062.
Martinez-Corral, I., Olmeda, D., Diéguez-Hurtado, R., Tammela, T., Alitalo, K., & Ortega, S. (2012). In vivo imaging of lymphatic vessels in development, wound healing, inflammation, and tumor metastasis. *Proceedings of the National Academy of Sciences of the United States of America, 109*, 6223–6228.
May, F., Hagedorn, I., Pleines, I., Bender, M., Vögtle, T., Eble, J., et al. (2009). CLEC-2 is an essential platelet-activating receptor in hemostasis and thrombosis. *Blood, 114*, 3464–3472.
Mertz, J. (2011). Optical sectioning microscopy with planar or structured illumination. *Nature Methods, 8*, 811–819.
Mounzer, R., Shkarin, P., Papademetris, X., Constable, T., Ruddle, N. H., & Fahmy, T. M. (2007). Dynamic imaging of lymphatic vessels and lymph nodes using a bimodal nanoparticulate contrast agent. *Lymphatic Research and Biology, 5*, 151–158.
Mumprecht, V., Honer, M., Vigl, B., Proulx, S. T., Trachsel, E., Kaspar, M., et al. (2010). In vivo imaging of inflammation- and tumor-induced lymph node lymphangiogenesis by immuno—Positron emission tomography. *Cancer Research, 70*, 8842–8851.
Niesner, R. A., Andresen, V., & Gunzer, M. (2008). Intravital two-photon microscopy: Focus on speed and time resolved imaging modalities. *Immunological Reviews, 221*, 7–25.
Ntziachristos, V. (2010). Going deeper than microscopy: The optical imaging frontier in biology. *Nature Methods, 7*, 603–614.
Oliver, G. (2004). Lymphatic vasculature development. *Nature Reviews Immunology, 4*, 35–45.
Oliver, G., & Detmar, M. (2002). The rediscovery of the lymphatic system: Old and new insights into the development and biological function of the lymphatic vasculature. *Genes & Development, 16*, 773–783.
Petrova, T. V., Makinen, T., Makela, T. P., Saarela, J., Virtanen, I., Ferrell, R. E., et al. (2002). Lymphatic endothelial reprogramming of vascular endothelial cells by the Prox-1 homeobox transcription factor. *EMBO Journal, 21*, 4593–4599.
Proulx, S., Luciani, P., Alitalo, A., Mumprecht, V., Christiansen, A., Huggenberger, R., et al. (2013). Non-invasive dynamic near-infrared imaging and quantification of vascular leakage in vivo. *Angiogenesis, 16*, 525–540.
Quintana, L., & Sharpe, J. (2011). Preparation of mouse embryos for optical projection tomography imaging. *Cold Spring Harbor Protocols, 2011*(6), 664–669.
Rasmussen, J. C., Tan, I. C., Marshall, M. V., Fife, C. E., & Sevick-Muraca, E. M. (2009). Lymphatic imaging in humans with near-infrared fluorescence. *Current Opinion in Biotechnology, 20*, 74–82.
Sabin, F. R. (1902). On the origin of the lymphatic system from the veins and the development of the lymph hearts and thoracic duct in the pig. *American Journal of Anatomy, 1*, 367–389.
Sabine, A., Agalarov, Y., Maby-El, H. H., Jaquet, M., Hagerling, R., Pollmann, C., et al. (2012). Mechanotransduction, PROX1, and FOXC2 cooperate to control connexin37 and calcineurin during lymphatic-valve formation. *Developmental Cell, 22*, 430–445.
Schacht, V., Ramirez, M. I., Hong, Y. K., Hirakawa, S., Feng, D., Harvey, N., et al. (2003). T1alpha/podoplanin deficiency disrupts normal lymphatic vasculature formation and causes lymphedema. *EMBO Journal, 22*, 3546–3556.
Schulte-Merker, S., Sabine, A., & Petrova, T. V. (2011). Lymphatic vascular morphogenesis in development, physiology, and disease. *Journal of Cell Biology, 193*, 607–618.

Sebzda, E., Hibbard, C., Sweeney, S., Abtahian, F., Bezman, N., Clemens, G., et al. (2006). Syk and Slp-76 mutant mice reveal a cell-autonomous hematopoietic cell contribution to vascular development. *Developmental Cell, 11*, 349–361.

Sevick-Muraca, E. M., Sharma, R., Rasmussen, J. C., Marshall, M. V., Wendt, J. A., Pham, H. Q., et al. (2008). Imaging of lymph flow in breast cancer patients after microdose administration of a near-infrared fluorophore: Feasibility study. *Radiology, 246*, 734–741.

Simons, M., & Eichmann, A. (2013). Lymphatics are in my veins. *Science, 341*, 622–624.

Srinivasan, R. S., Geng, X., Yang, Y., Wang, Y., Mukatira, S., Studer, M., et al. (2010). The nuclear hormone receptor Coup-TFII is required for the initiation and early maintenance of Prox1 expression in lymphatic endothelial cells. *Genes and Development, 24*, 696–707.

Srinivasan, R. S., & Oliver, G. (2011). Prox1 dosage controls the number of lymphatic endothelial cell progenitors and the formation of the lymphovenous valves. *Genes and Development, 25*, 2187–2197.

Steven, P., Bock, F., Huttmann, G., & Cursiefen, C. (2011). Intravital two-photon microscopy of immune cell dynamics in corneal lymphatic vessels. *PLoS One, 6*, e26253.

Suzuki-Inoue, K., Fuller, G. L. J., García, Á., Eble, J. A., Pöhlmann, S., Inoue, O., et al. (2006). A novel Syk-dependent mechanism of platelet activation by the C-type lectin receptor CLEC-2. *Blood, 107*, 542–549.

Suzuki-Inoue, K., Inoue, O., Ding, G., Nishimura, S., Hokamura, K., Eto, K., et al. (2010). Essential in vivo roles of the C-type lectin receptor CLEC-2: Embryonic/neonatal lethality of CLEC-2-deficient mice by blood/lymphatic misconnections and impaired thrombus formation of CLEC-2-deficient platelets. *Journal of Biological Chemistry, 285*, 24494–24507.

Tammela, T., & Alitalo, K. (2010). Lymphangiogenesis: Molecular mechanisms and future promise. *Cell, 140*, 460–476.

Truman, L. A., Bentley, K. L., Smith, E. C., Massaro, S. A., Gonzalez, D. G., Haberman, A. M., et al. (2012). ProxTom lymphatic vessel reporter mice reveal Prox1 expression in the adrenal medulla, megakaryocytes, and platelets. *American Journal of Pathology, 180*, 1715–1725.

Uhrin, P., Zaujec, J., Breuss, J. M., Olcaydu, D., Chrenek, P., Stockinger, H., et al. (2010). Novel function for blood platelets and podoplanin in developmental separation of blood and lymphatic circulation. *Blood, 115*, 3997–4005.

Vakoc, B. J., Fukumura, D., Jain, R. K., & Bouma, B. E. (2012). Cancer imaging by optical coherence tomography: Preclinical progress and clinical potential. *Nature Reviews Cancer, 12*, 363–368.

Van Balkom, I. D. C., Alders, M., Allanson, J., Bellini, C., Frank, U., De Jong, G., et al. (2002). Lymphedema—lymphangiectasia—mental retardation (Hennekam) syndrome: A review. *American Journal of Medical Genetics, 112*, 412–421.

Wigle, J. T., & Oliver, G. (1999). Prox1 function is required for the development of the murine lymphatic system. *Cell, 98*, 769–778.

Yang, Y., Garcia-Verdugo, J. M., Soriano-Navarro, M., Srinivasan, R. S., Scallan, J. P., Singh, M. K., et al. (2012). Lymphatic endothelial progenitors bud from the cardinal vein and intersomitic vessels in mammalian embryos. *Blood, 120*, 2340–2348.

Yuen, D., Wu, X., Kwan, A. C., LeDue, J., Zhang, H., Ecoiffier, T., et al. (2011). Live imaging of newly formed lymphatic vessels in the cornea. *Cell Research, 21*, 1745–1749.

Zeeb, M., Axnick, J., Planas-Paz, L., Hartmann, T., Strilic, B., & Lammert, E. (2012). Pharmacological manipulation of blood and lymphatic vascularization in ex vivo-cultured mouse embryos. *Nature Protocols, 7*, 1970–1982.

Chapter 14
Clinical Disorders of Primary Malfunctioning of the Lymphatic System

Carlo Bellini and Raoul CM Hennekam

Abstract Primary lymphedema is defined as lymphedema caused by dysplasia of the lymph vessels. This complex group of diseases is discussed in detail from a clinical perspective. A review of the epidemiology and classification of lymphedema on the backdrop of its clinical presentation reveals weaknesses of the present classification system, which, however, is the basis for the choice of optimal patient care. Non-syndrome and syndrome types of primary lymphedema are presented in detail and related molecular findings are summarized.

14.1 Introduction

Lymphedema is a chronic, often progressive swelling of subcutaneous tissue due to failure of the lymphatic system to drain fluid from the interstitial spaces, causing fluid accumulation. Clinically a distinction is made between "primary" and "secondary" lymphedema (Rockson and Rivera 2008). Primary lymphedema is defined as lymphedema caused by dysplasia of the lymph vessels. It is usually congenital and genetically determined. It can be either isolated, so without manifestations in other tissues or outside the lymph vessels, or be part of a disorder that shows other signs and/or symptoms as well (syndrome). The distinction between isolated forms of lymphedema and those that are part of a more generalized entity is not strict as it also depends on the detail of the studies in affected individuals to search for other characteristics next to lymphedema. For instance the presence of an additional row of eyelashes (distichiasis) can be easily missed if not specifically searched for. Primary lymphedema usually affects the extremities as a result of abnormal

C. Bellini
Neonatal Intensive Care Unit, Emergency Department, Gaslini Institute, Genoa, Italy

R.CM. Hennekam (✉)
Department of Pediatrics and Translational Genetics, Academic Medical Center, Amsterdam, The Netherlands
e-mail: r.c.hennekam@amc.uva.nl

F. Kiefer and S. Schulte-Merker (eds.), *Developmental Aspects of the Lymphatic Vascular System*, Advances in Anatomy, Embryology and Cell Biology 214,
DOI 10.1007/978-3-7091-1646-3_14,

regional lymph drainage, although visceral drainage showing in lymphangiectasias of for instance gut or lung can also be impaired. Secondary lymphedema is acquired, typically as consequence of an infection, trauma, or malignancy, and will not be discussed any further in this chapter.

Primary lymphedema in children can cause considerable diagnostic difficulties to clinicians and distress to parents. It is essential to obtain a rapid diagnosis and to implement correct treatment at the earliest opportunity. It is estimated that many physicians and surgeons will see less than ten cases of lymphedema in a year (Tiwari et al. 2006). It is therefore imperative that patients are referred at an early stage to a clinic with wide experience and expertise in diagnostics and treatment. Primary lymphedema can also show in lymphangiectasia of internal organs. When affecting the intestines it produces a protein-losing enteropathy and severe malabsorption of lipids and other nutrients (Braamskamp et al. 2010) Congenital pulmonary lymphangiectasia is a rare developmental disorder involving the lung and characterized by pulmonary subpleural, interlobar, perivascular, and peribronchial lymphatic dilatation, complicated by chylous pleural effusion (Bellini et al. 2006). Lymphangiectasias can also occur in other internal organs such as the pericardium, kidneys, and thyroid gland (Van Balkom et al. 2002). Lymphedema should be discerned from lipedema. This is a poorly understood condition characterized by swelling and enlargement of the lower limbs due to abnormal deposition of subcutaneous fat. Lymphatic system involvement seems likely but is actually debated (Child et al. 2010).

The increased knowledge regarding the etiology and pathogenesis of inherited disorders involving the lymphatic system has offered further insight in lymph vessel formation in general. Developments in lymphatic biology and various pathways and mechanisms through which the lymphatic system contributes to the pathogenesis of disorders have been reviewed elsewhere extensively (Tammela and Alitalo 2010; Alitalo 2011; Martinez-Corral and Makinen 2013) and are out of the scope of the present chapter.

14.2 Epidemiology

According to World Health Organization, lymphedema has a worldwide incidence of 300 million cases (~1 in every 20 individuals). Almost half of lymphedemas are of primary origin, due to congenital lymphatic dysplasia and subsequent poor functioning of lymph nodes and/or lymphatic vessels. Some 70 million are of parasitic origin (especially Filaria Bancrofti); 50 million are postsurgery cases, often following breast cancer surgery. The remaining 30 million cases are likely caused by functional problems related to water overload on lymphatic circulation (http://www.chirurgiadeilinfatici.it/en/lymphatic-diseases/lymphedema/epidemiology). The exact prevalence of primary lymphedema is unknown. Within the USA, it has been estimated to be 1.15 per 100,000 children (Smeltzer et al. 1985). A population prevalence of 1.33 per 1,000 for all ages has been reported, but it is probably an underestimation of the true burden of disease (Moffatt et al. 2003). A female preponderance (M:F = 1:3) is documented, although in part this may represent ascertainment bias.

14.3 Classifications

Primary lymphedema is chronic edema, in which fluid accumulates due to abnormal structure or functions of the lymphatic system (Mortimer 1995). In most cases, edema will be present from birth, but in some cases the lymphedema develops at a later age despite the lymphatic dysplasia being present congenitally. Possibly an increased need of lymphatic functioning due to increased body size and weight or other factors such as hormones or external influences plays a role. Primary isolated lymphatic dysplasias constitute a spectrum of disorders that may manifest by a variety of clinical presentations: lymphedema, chylous effusions, lymphangiomatous malformations with cystic masses and localized gigantism, intestinal lymphangiectasia with malabsorption, lung lymphangiectasia, and lymphangiectasias of other internal organs and glands. Clinical classification of the various types of primary isolated lymphedema has historically been into three groups: lymphedema congenita, lymphedema praecox, and lymphedema tarda. Such classification based purely on the age of onset of the lymphedema does not take into account many other aspects of lymphedema and hinders a refined, detailed classification of phenotypes.

Anomalies of the lymphatic system should be considered as part of vascular anomalies. Mulliken and Glowacki (1982) proposed a classification system for vascular anomalies based on the clinical manifestations and endothelial cell characteristics into two main groups, i.e., hemangiomas and vascular malformations. This classification was adopted by the International Society for the Study of Vascular Anomalies (ISSVA) and with some subsequent modifications widely accepted. At present the two main types of vascular anomalies are vascular tumors (the most common type being hemangioma) and vascular malformations (Enjolras and Mulliken 1997) (Table 14.1). The term malformation that is used in this and other publications to indicate this is mostly wrong as the anomalies are in fact no malformations but dysplasias, although exceptions exist (Hennekam et al. 2013). However, to avoid confusion, we have chosen to use here the terminology used by the ISSVA and will discuss the right terminology with the ISSVA. Vascular malformations include slow-flow malformations which contain the lymphatic malformations (LM). The ISSVA uses a similar classification, splitting the congenital lymphatic dysplasias in truncular (T) or extratruncular (ET), depending on endothelial cells characteristics and the embryonic stage at which the defect was produced. The ET-LMs are embryonic remnants, which have occurred during early stages of vasculogenesis. The immature mesenchymal tissue (in fact dysplastic tissue), of which these malformations are formed, maintains a proliferative capacity. In the case stimuli occur such as pregnancy, hormonal stimulation, trauma, or

Table 14.1 Main characteristics of primary isolated lymphatic malfunctioning

Slow-flow malformations
Divided into truncular or extratruncular lesions
Extratruncular lesions are cystic and divided into microcystic, macrocystic, and mixed type
Extratruncular lesions maintain a proliferative capacity
Truncular lesions are linked to primary lymphedema, lymphangiectasia, and lymphangiomatosis
Truncular lesions have no proliferative capacity but behave as malformations
Extratruncular and truncular lesions may coexist within a single patient

surgery, ET-LMs can be stimulated and develop into (micro- and/or macro-) cystic lesions: lymphangioma. The T-LMs develop later during embryogenesis. The vascular tissue is mature and no longer has proliferative capacity. Primary lymphedema is linked with such malformations (Lee et al. 2005).

The Hamburg classification (7th international ISSVA workshop on vascular anomalies, Hamburg, 1988) distinguishes congenital vascular malformations (CVMs) in truncular (T) and extratruncular (ET) forms (Belov 1993; Lee et al. 2005). The Hamburg classification includes a further group of mixed venous malformations (VMs), identified as hemolymphatic malformation (HLM), making the classification of lymphatic malformations (LM) difficult and confusing. The majority of LM lesion exists as an "independent" form of the CVM, either as primary lymphedema representing "truncular" LM lesion or as cystic, cavernous, or capillary lymphangioma representing "extratruncular" LM lesion. Extratruncular LM lesion and truncular LM lesion co-occur together infrequently. In our opinion the Hamburg classification causes considerable problems. It is often extremely difficult to determine whether a finding is truncular or extratruncular as in fact this asks for detailed embryological studies in animal models and in fact also in humans, and the subdivision as is made now is build only for a small part on solid embryological grounds. The various terms used in the classification are often not sufficiently carefully chosen as malformations and dysplasias are insufficiently discerned from one another while this distinction has significant consequences in patient care. Lastly, the subdivision of the various forms of lymphatic malfunctioning is in our opinion not helpful in diagnostics or in providing optimal care to patients, and these issues should be the main determinants in any classification of disorders. A new classification of lymphatic malfunctioning is urgently needed.

Clinically, lymph reflux (backflow of lymph) may occur localized or at systemic level. The backflow of chyle from intestine is characteristic of intestinal lymphangiectasia. Intestinal lymphangiectasia results in protein-losing gastroenteropathy, but it can also affect the abdominal lymphatics or the thoracic duct; if the latter occurs, a chylothorax may develop. The primitive backflow at the level of pleural and pulmonary lymphatics results in pulmonary lymphangiectasia. Lymphangiectasia presents as dilated lymphatics and is usually associated with lymphedema. The pressure in the dilated lymphatics is increased, causing leakage of lymph into surrounding tissues. Lymphangiomatosis is characterized by well-differentiated lymphatic capillaries which are dilated forming cysts and are not always associated with lymphedema. This collection of dilated lymphatics is typically isolated from the remainder of the lymphatic system, which can be completely normal. These are thought to arise from inappropriate connection of the embryonic lymph sacs with the lymphatic system during embryogenesis. The lymphangioma may be uni- or multilocular and macro- or microcystic and may occur in any part of the body. The most common site is the neck (cystic hygroma). Lymphangiomatosis is the widespread, multifocal occurrence of lymphangioma, which can grow aggressively. Lymphangiomatosis can be difficult to diagnose.

In general, LM may coexist with a wide spectrum of CVM and thus be part of complex disorders, affecting the entire circulation system: arteries, veins, lymphatics, and capillaries. This occurrence can be demonstrated in Klippel-Trenaunay syndrome, in which LM may coexist with venous and capillary malformation, or in Parkes Weber syndrome, in which LM presents in association with AV malformation. Indeed, in the entity that goes along with the most widespread lymphatic dysplasia, Hennekam syndrome, the co-occurrence of anomalies of other parts of the vascular system has been described (Van Balkom et al. 2002; Alders et al. 2013). Table 14.1 summarizes the main characteristics of the LMs.

14.4 Isolated Types

Primary congenital lymphedema (*Milroy syndrome*) is an autosomal dominant disorder of the peripheral lymphatics characterized by lower limb lymphedema, typically affecting the dorsum of the feet. It is usually bilateral and present at birth or evident soon thereafter. Milroy syndrome can also present as lymphedema of the upper limbs, or in markedly affected individuals, the lymphedema can start at the lower limbs and become present at the upper limbs later in life. In such individuals also the external genitalia may become affected, and the differentiation with more marked lymphedema as can be present in Hennekam syndrome may be difficult. Indeed molecularly proven cases with chylothorax and hydrops fetalis have been described (Daniel-Spiegel et al. 2005). Usually the lymphedema in Milroy syndrome becomes gradually more marked during life although rarely it can improve during life as well. The severity of lymphedema shows a marked variability, also intrafamilial, and careful evaluation of family members is regularly needed to establish whom in the family is affected or not. Milroy syndrome can be caused by mutations in *FLT4* and *GJC2* and may also be caused by *VEGFC* mutations.

Primary lymphedema at an older age (*Meige syndrome*) is an autosomal dominant disorder characterized by peripheral lymphedema predominantly in the lower limbs with onset around puberty. It is thought that the lymphatic system normally functions at ~10 % capacity (Connell et al. 2009). It is assumed to be caused by underdevelopment of the lymphatic vessels, which is however still sufficient in the first years of life but becomes functionally insufficient with increased body size and due to other such as puberty. Upper limb and facial involvement can also present.

The lymphedema, which occurs in Meige syndrome, is clinically indistinguishable from that found in the lymphedema-distichiasis syndrome. Indeed there have been publications reporting on *FOXC2* mutations in Meige syndrome, but likely this was a family with lymphedema-distichiasis syndrome. Until now no causative gene for Meige syndrome has been reported. It should be noted there is also an adult-onset segmental dystonia that is termed Meige syndrome (OMIM #128100).

Primary intestinal lymphangiectasia (Vignes and Bellanger 2008) is a disorder with unknown prevalence but which seems to occur only infrequently. It is characterized by hypoproteinemia, edema, and lymphocytopenia, resulting from loss of lymph fluid into the gastrointestinal tract due to intestinal lymphatic vessels dilatation, thus resulting in protein-losing gastro-enteropathy. The loss of lymph fluid can be confirmed by the elevated 24-h clearance of alpha-1-antitrypsin in stools. Bilateral lower limb edema and diarrhea are typical clinical signs that are secondary to the gastro-enteropathy. Lymphocytopenia, hypogammaglobulinemia, hypocalcemia, trace metal deficiency due to malabsorption, and chylous pleural effusions and chylous ascites may occur as secondary consequences in long-standing lymphangiectasia. Primary limb lymphedema may be present as well which can be difficult to distinguish from edema. If this occurs, more generalized lymphatic dysplasia such as occurs in Hennekam syndrome must be considered.

Primary pulmonary lymphangiectasia (Bellini et al. 2006) is an infrequently described developmental disorder involving the lung and characterized by pulmonary subpleural, interlobar, perivascular, and peribronchial lymphatic dilatation. The prevalence is unknown. Pulmonary lymphangiectasias typically present at birth with severe respiratory distress, tachypnea, and cyanosis, with a very high mortality rate at birth or within a few hours of birth. Most reported cases are sporadic and the etiology remains unexplained. Patients affected by PL who survive infancy present medical problems which are characteristic of chronic lung disease. Pulmonary lymphangiectasias also develop at a later age (often puberty of adolescence) in more generalized lymphatic dysplasias such as Hennekam syndrome. Also in such individuals the course is typically unpredictable but eventually fatal.

Chylothorax is defined as an accumulation of chyle in the pleural space. Chylothorax should be considered as a common endpoint for a variety of pathological processes including intrinsic abnormalities of the lymphatic system or disruption of the thoracic duct via trauma, surgery, malignancy, or cardiovascular disease. Congenital defects of the thoracic duct, either isolated or associated with generalized lymphatic vessel dysplasia, are the most frequent cause of congenital chylothorax. Congenital chylothorax is a rare cause of respiratory distress in the newborn but is the most common form of pleural effusion in the neonatal period. Reported incidence ranges from 1:1,000 to 1:15,000 pregnancies (Dubin et al. 2000). The actual incidence in man is probably higher, as intrauterine fetal death as well as stillbirth might well be underestimated. Although familial occurrence has been reported the exact pattern inheritance is not yet known. There is a 2:1 male to female predominance. Both X-linked and autosomal recessive inheritance have been suggested (Straats et al. 1980) and in our opinion it is likely genetically heterogeneous. It has been suggested that congenital pulmonary lymphangiectasia is a constant pathological finding in congenital chylothorax and that this may imply a common pathogenesis for these disorders (Bellini et al., 2006; Bellini et al., in press).

The course of congenital chylothorax varies widely and the prognosis is unpredictable. Overall mortality for congenital chylothorax has been reported as high as 50 %. The presence of hydrops fetalis has significant prognostic implications (Dubin et al. 2000). In cases of chylothorax complicated by hydrops fetalis, a decrease in survival from 100 to 52 % has been reported. Still, nonimmune hydrops caused by chylothorax carries a better prognosis than nonimmune hydrops in general. The frequency of spontaneous resolution, which may occur either before or after birth, is still unknown. Lymphangiomatosis and lymphangiectasia are the two main anomalies of lymphatic development that cause chylothorax (Fox et al. 1998).

14.5 Syndrome Types

We performed a literature search to obtain an overview of syndromic forms of primary lymphatic malfunctioning. We used the online edition of Mendelian Inheritance in Man and the Winter-Baraitser Dysmorphology Database (WBDD), using as search terms lymphedema, lymphangiectasia, and chylothorax. OMIM is the comprehensive compendium of human genes and genetically determined phenotypes. OMIM contains information on all known Mendelian disorders and over 12,000 genes (http://www.ncbi.nlm.nih.gov/omim). The Winter–Baraitser Dysmorphology Database currently contains information on ~6,000 entities characterized by one or more morphologic abnormalities. It includes Mendelian disorders, chromosomal imbalances, sporadic conditions, and those caused by environmental agents (http://www.lmdatabases.com/).

The thus retrieved entries are summarized in Table 14.2, which contains the main characteristics of each entity such as chromosomal locus, gene involved, pattern of inheritance, and major clinical manifestations. It is impossible to discuss each entity listed in detail. Therefore we will only provide short descriptions of three entities that we consider paradigmatic of the various form of lymphatic maldevelopment.

Lymphedema-distichiasis syndrome is a single gene disorder caused by *FOXC2* (forkhead transcription factor) mutations (Sutkowska et al. 2012). Distichiasis (from Greek "distikhos," meaning two rows) is a congenital anomalous growth of eyelashes from the meibomian glands of the eyelid, causing the presence of a double row of eyelashes. It has been suggested that the lymphatic vessel malfunction may be linked to lymphatic valvular insufficiency which causes marked lymphatic reflux (Brice et al. 2002). The severity of lymphedema varies among families and among affected individuals of a single family and is linked to the grade of lymph reflux. "Yellow nail syndrome" has been published as a separate entity characterized by lymphedema and yellow, dystrophic, thick, and slowly growing

Table 14.2 Main characteristics of primary lymphatic malfunctioning as part of a syndrome

Syndrome	Pattern of inheritance	Prevalence	Location	Gene	OMIM number	Orphanet identifier	Clinical synopsis
Aagenaes (recurrent cholestasis, lymphedema)	AR AD	Rare	15q	?	214,900	ORPHA1414	Chronic lymphedema, recurrent neonatal cholestasis
Al-Gazali-Bakalinova (macrocephaly, multiple epiphyseal dysplasia)	AR	Rare	15q26.1	KIF7	607,131	ORPHA166024	Macrocephaly, multiple epiphyseal dysplasia, unusual face, lymphedema distal limbs
Amor (Adams–Oliver-like)	AR	Rare	19p13	DOCK6	614,219	ORPHA974	Intellectual disability, transverse limb anomalies, aplasia cutis scalp, cortical dysplasia, limb lymphedema
Avasthey (pulmonary hypertension, cranial arteriovenous malformations, lymphedema)	AD	Rare	?	?	152,900	OPHA86914	Cranial av malformations, pulmonary hypertension, childhood distal limb lymphedema
Bronspiegel (aplasia cutis congenita, intestinal lymphangiectasia)	AR	Rare	?	?	207,731	ORPHA1116	Aplasia cutis scalp, coloboma optic disk, heterotopia, intestinal lymphangiectasia
Cantu (hypertrichosis, osteodysplasia, cardiomyopathy)	AD	Unknown	12p12	ABCC9	239,850	ORPHA1517	Hypertrichosis, mildly coarse face, wide ribs, flat vertebrae, pericardial effusion, cardiomyopathy, limb lymphedema in adulthood
Cerebellar hypoplasia, lissencephaly, lymphedema	AR	Rare	7q22	RELN	257,320	ORPHA89844	Lissencephaly, seizures, hypotonia, ataxia, small cerebellum, limb lymphedema
Choanal atresia-lymphedema	AR	Rare	1q41	PTPN14	613,611	–	Choanal atresia, small nipples, childhood limb lymphedema

Chromosome 5p13 duplication	Chromosomal	Rare	5p13	?	–	–	Intellectual disability, obesity, macrocephaly, hypertelorism, loose skin neck, lymphedema
Chromosome 6q27 deletion	Chromosomal	Rare	6q27	?	–	–	Intellectual disability, autism, seizures, small callosal body, pulmonary lymphangiectasia
Chromosome 8q24 deletion	Chromosomal	Rare	8q24	?	–	–	Intellectual disability, sparse hair, broad nose, fractures, brachydactyly, limb lymphedema
Costello	AD	Unknown	11p15.5 12p12.1	HRAS KRAS	218,040	ORPHA3071	Intellectual disability, short stature, curly hair, coarse face, nasal papillomata, cardiomyopathy, pulmonary lymphangiectasia
Dahlberg (hypoparathyroidism, lymphedema)	AR/XL?	Rare	?	?	247,410	ORPHA1563	Hypoparathyroidism, ptosis, telecanthi, nephropathy, brachytelephalangy, congenital limb lymphedema,
Da Silva Lopes (frontonasal dysplasia, neuronal migration defect, lymphedema)	?	Rare	?	?	136,760	ORPHA250	Intellectual disability, cortical dysplasia, broad forehead, hypertelorism, congenital limb lymphedema
Distichiasis-lymphedema	AD	Unknown	16q24.1	FOXC2	153,400	ORPHA33001	Double row of eyelashes, ptosis, cleft palate, limb lymphedema in puberty
Ectodermal dysplasia, immune deficiency (OLEDAID)	XL	1:250,000 males	Xq28	IKBKG	300,301	ORPHA98813	Sparse hair, oligodontia, dry skin, immunodeficiency, osteopetrosis, congenital lymphedema
Emberger (myelodysplasia, lymphedema)	AD	Rare	3q21.3	GATA2	614,038	ORPHA3226	Myelodysplasia, leukemia, deafness, lymphedema of lower limbs/genitalia in infancy-puberty

(continued)

Table 14.2 (continued)

Syndrome	Pattern of inheritance	Prevalence	Location	Gene	OMIM number	Orphanet identifier	Clinical synopsis
German (hypotonia, arthrogryposis, unusual face, lymphedema)	AR	Rare	?	?	231,080	ORPHA2077	Hypotonia, arthrogryposis, metopic ridge, cleft palate, hand lymphedema
Gilewski (unusual face, eventration diaphragm, pulmonary lymphangiectasia)	?	Rare	?	?	–	–	Broad nasal bridge, small nose, small mouth, eventration diaphragm, small penis, pulmonary lymphangiectasia
Hennekam (intestinal lymphangiectasia, lymphedema, intellectual disability)	AR	Unknown	18q21.32	CCBE1	235,510	ORPHA2136	Intellectual disability, flat face, hypertelorism, flat nasal bridge, glaucoma, intestinal lymphangiectasia, congenital severe lymphedema of limbs, genitalia, and face
Hennekam (epidermal nevus, arteriovenous malformations, intestinal lymphangiectasia)	?	Rare	?	?	–	–	Epidermal nevus, lipoma, intramedullary hemangioma, intestinal lymphangiectasia
Hypotrichosis, lymphedema, telangiectasia	AR	Rare	20q13.33	SOX18	607,823	ORPHA69735	Progressive hypotrichosis, telangiectasia, childhood-pubertal lower limb lymphedema
Irons-Bianchi (congenital heart anomaly, unusual face, lymphedema)	AR	Rare	?	?	601,927	ORPHA86915	Congenital heart defects, prominent forehead, flat nasal bridge, congenital limb lymphedema
Jaeken (CDGIa)	AR	1/20,000	16p13.3	PMM2	212,065	ORPHA79318	Intellectual disability, small cerebellum, retinitis pigmentosa, abnormal fat pads, stroke-like episodes cardiomyopathy, congenital limb lymphedema

Klippel-Trenaunay	AD	1/25,000	?	?	149,000	ORPHA90308	Cutaneous hemangioma, arteriovenous malformations, lymphangioma, asymmetry, childhood-adulthood lymphedema
Lipedema	AD	Unknown	?	?	614,103	ORPHA77243	Fat legs, orthostatic edema, lower limb lymphedema
Lymphedema, agenesis corpus callosum	AR	Rare	?	?	613,623	–	Intellectual disability, callosal body agenesis, congenital resolving limb lymphedema
Mandibulofacial dysostosis, polydactyly, lymphedema	?	Rare	?	?	–	–	Underdeveloped malae, small nose, small ears, deafness, postaxial polydactyly hands and feet, congenital limb lymphedema
Meige	AD	Unknown	?	?	153,200	ORPHA90186	Pubertal lymphedema in lower limbs
Microcephaly-capillary malformation	AR	Unknown	2p13	STAMBP	614,261	ORPHA294016	Intellectual disability, microcephaly, cortical dysplasia, small distal phalanges, multiple capillary malformations, limb lymphedema
Microcephaly, cutis verticis gyrata, lymphedema	AR	Rare	?	?	–	–	Intellectual disability, cutis gyrate scalp, microcephaly, enlarged liver and spleen, congenital lymphedema
Microcephaly, chorioretinal dysplasia, lymphedema	AD	Rare	10q23.33	KIF11	152,950	ORPHA2526	Intellectual disability, microcephaly, chorioretinal pigmentation, congenital distal limb lymphedema
Milroy	AD	Unknown	5q35.3 1q42.13 4q34.3 6q16.2	FLT4 GJC2 VEGFC	153,100 613,480 601528	ORPHA79452	Congenital limb/genitalia lymphedema

(continued)

Table 14.2 (continued)

Syndrome	Pattern of inheritance	Prevalence	Location	Gene	OMIM number	Orphanet identifier	Clinical synopsis
Njolstad (pulmonary lymphangiectasia, lymphedema)	AR	Rare	?	?	265,300	ORPHA2414	Flat face, hypertelorism, flat nasal bridge, glaucoma, congenital pulmonary lymphangiectasia, congenital lymphedema
Noonan	AD	1/2,500	12q24.13 12p12.1 2p22.1 3p25.2 1p13.2 7q34 1q22	PTPN11 KRAS SOS1 RAF1 NRAS BRAF RIT1	163,950 190,070 610,733 611,552 613,224 613,706 615,355	ORPHA648 ORPHA98733	Short stature, unusual face, webbed neck, pulmonic stenosis, pectus carinatum, intestinal lymphangiectasia, limb lymphedema
Opitz Frias	XL Chromosomal	Rare	Xp22.2 22q11.2	MID1 deletion	145,410	ORPHA2745	Intellectual disability, hypertelorism, cleft palate, laryngeal cleft, hypospadias, pulmonary lymphangiectasia
PTEN hamartoma tumor	AD	1/200,000	10q23.31	PTEN	153,480	ORPHA109	Macrocephaly, hemangioma, lipoma, tumors, pulmonary lymphangiectasia
Pulmonary stenosis, lymphedema	XL	Rare	Xp11	?	–	–	Peripheral pulmonary stenosis, distal limb lymphedema
Prolidase deficiency	AR	Rare	19q13.11	PEPD	170,100	ORPHA742	Intellectual disability, short stature, photosensitivity, recurrent limb ulcers, childhood lymphedema
Schindler disease (alpha-*N*-acetylgalactosaminidase deficiency)	AR	Rare	22q13.2	NAGA	104,170	–	Intellectual disability, cortical blindness, deafness, seizures, limb lymphedema

Silver (acro-osteolysis, intestinal lymphangiectasia)	?	Rare	?	?	–	–	Clubbing, acro-osteolysis, bowed radius and ulna, intestinal lymphangiectasia
Stoll (brachydactyly, tachycardia, lymphedema)	AD	Rare	?	?	–	–	Brachydactyly, syndactyly paroxysmal tachycardia, pubertal limb lymphedema
Tuberous sclerosis	AD	1/10,000	9q34.13 16p13.3	TSC1 TSC2	191,100 613,254	ORPHA805	Depigmented skin lesions, angiofibroma, hamartoma in heart/kidneys/lung/brain, congenital lymphedema
Turner	Chromosomal	1/1,000	X	?	–	ORPHA881	Short stature, ovarian failure, variable visceral manifestations, congenital limb and sometimes facial lymphedema
Urioste	AR	Rare	?	?	235,255	ORPHA1655	Oligodactyly, polydactyly, vertebral anomalies, aplasia of internal organs, congenital intestinal lymphangiectasia
Yellow nail	AD	Unknown	16q24.1	FOXC2	153,300	ORPHA662	Yellow nails, chylothorax, congenital adulthood limb lymphedema
Zellweger	AR	1/25,000–50,000	7q21.2	PEX1	214,100	ORPHA912	Intellectual disability, hypotonia, seizures, unusual face, enlarged liver, intestinal lymphangiectasia

AD autosomal dominant, *AR* autosomal recessive, *XL* X-linked recessive, *?* not known

nails. However, it has become likely that the thickening and yellow discoloration of the nails is not a distinctive sign and can occur in several marked forms of distal limb lymphedema, and most patients with "yellow nail syndrome" may in fact have had lymphedema-distichiasis syndrome (Rezaie et al. 2008).

Hennekam syndrome is a form of very marked lymphatic dysplasia, in which lymph vessels in all body areas are affected (Van Balkom et al. 2002; Alders et al. 2013) and which is caused by mutation in *CCBE1* (Alders et al. 2009). These are also already affected prenatally, and the facial manifestations are thought to be explainable this way. According to the original description (Hennekam et al. 1989) the main characteristics are lymphedema, intestinal lymphangiectasia, intellectual disability (which can be markedly different, also within a single sibship, varying from moderate to severe intellectual disability to completely normal development), and facial signs. The lymphedema has always been congenital, sometimes markedly asymmetric, and after initial decrease in the first years of life has become often gradually progressive with age. Lymphangiectasias are not limited to the intestines but can also be found in the lungs, pleura, pericardium, thyroid gland, and kidneys.

Noonan syndrome is a well-known and frequent entity that is mainly characterized by short stature, unusual face, webbing of the neck, a combined pectus carinatum and excavatum, pulmonic stenosis and later on an increased chance to develop a cardiomyopathy, and a host of further major and minor anomalies. Congenital lymphedema of the distal upper and lower limbs is often but not always present. Some infants and children develop more marked lymphatic malfunctioning including intestinal lymphangiectasias and chylothorax. Within families the variability of the lymphatic system involvement can vary very widely. Indeed even newborns with fatal hydrops have been born to an affected parent with only limited manifestations of the syndrome. Noonan syndrome can be caused by a series of genes that all act in the same MAPK pathway and is one of the entities that form the rasopathies (Rauen 2013).

14.6 Molecular Findings

In isolated and syndromic primary lymphedema various patterns of inheritance can be recognized. The etiology and pathogenesis of the group of disorders is only partially understood (Table 14.2), but research in this field has improved significantly in recent years due to advances in sequencing techniques. Mutations in *VEGFR3*, *FOXC2*, and *SOX18* are known to cause Milroy disease, lymphedema-distichiasis syndrome, and hypotrichosis-telangiectasia-lymphedema syndrome, respectively (Ferrell et al. 1998; Fang et al. 2000; Irrthum et al. 2003; Brice et al. 2005) and *VEGFC* mutations have been found in a Milroy-like disorder (Gordon et al. 2013). *CCBE1* has been reported to be mutated in a proportion of patients with Hennekam syndrome and the analysis of this gene should be studied in every patient with a Hennekam syndrome phenotype or otherwise marked lymphatic malfunctioning (Alders et al. 2009, 2013; Connell et al. 2010), irrespective of the cognitive functioning of affected individuals or the presence or absence of other abnormalities. Chromosome imbalances often result in multiple organ defects and the lymphatic system can be part of this as well, although this is not common, except for Turner syndrome. The number of genes known to cause isolated primary lymphatic malfunctioning is still relatively small compared to the number of genes known to cause similar vascular malfunctioning, and we may expect several other genes to be recognized as being involved in lymphangiogenesis or lymphatic functioning. Indeed sequencing of a series of biologically plausible candidate genes such as *PROX1*, *EMILLIN1*, *LCP2*, *LYVE1*, *NRP2*, *PDPN*, and *SYK* has been suggested as these may be involved in primary lymphedema families.

14.7 Diagnostic Work-Up

The approach to establish the diagnosis in the often complex and sometimes confusing lymphatic disorders can cause difficulties. We suggest a general scheme to provide help in the diagnostic process that can be generally applied in disorders that go along with lymphatic malfunctioning. In all patients a detailed family history of at least two generations and including information to the existence of a possible consanguinity between the parents should be obtained. The examination of a fetus with a suspected lymphatic dysplasia may comprise amniotic fluid examination and chorionic villi study, in particular searching for inherited metabolic disease, lysosomal storage diseases (LSD) included. The examination of a deceased fetus may include autopsy, including babygram, photo-documentation, immunohistochemical studies, and also a skin biopsy (both for DNA collection and to have access to cultivated cells), samples of other tissues, examination of the placenta, and obtaining and evaluation of fetal urine for metabolic disorders.

Lymphedema in an infant or child is usually diagnosed clinically, but in case of doubt lymphoscintigraphy is the main instrument to establish the diagnosis of lymphedema and to visualize peripheral lymphatic vessels. Lymphoscintigraphy relies on one of the essential functions of the lymphatic system, i.e., to transport lymph. Thus, it can demonstrate each defect in lymphatic functions, as delayed lymphatic drainage, asymmetric or absent visualization of regional lymph nodes, dermal backflow, and interrupted lymphatic structures (Keeley 2006). It can study both superficial and deep lymphatic circulation. Lymphoscintigraphy has been demonstrated to be safe and effective in newborns and in children (Bellini et al. 2008). The diagnostic approach in an individual with visceral lymphangiectasia is usually complex. Lymphoscintigraphy may be helpful and is often combined with computerized tomography scanning and interstitial magnetic resonance imaging. These combined methods of studying lymphatic anomalies are nor widely available however. Additional laboratory exams are usually indicated if lymphangiectasias are present. Determining fecal excretion of alpha-1-antitrypsin helps in diagnosing a protein-losing syndrome and points to intestinal lymphangiectasias. If visceral effusions are present, such as chylothorax, chylopericardium, and chylous ascites, the abnormal fluids should be obtained and examined to evaluate the triglyceride level (pointing to lymph if >110 mg/dL [= 1.1 mmol/L]), presence of chylomicrons (Sudan III staining), determining the absolute cell count (pointing to lymph if >1,000 cs/mcL, with a lymphocyte fraction of >75–90 %), and the cholesterol level (abnormal if 60 mg/dL or above) (Bellini et al. 2012).

Recognizing congenital lymphatic disorders and diagnosing reliably lymphatic malfunctioning are still challenging. It still starts with clinical recognition, but likely other tools that are patient-friendly such as molecular genetics and other biomarkers can be developed in the near future.

References

Alders, M., Hogan, B. M., Gjini, E., Salehi, F., Al-Gazali, L., Hennekam, E. A., et al. (2009). Mutations in CCBE1 cause generalized lymph vessel dysplasia in humans. *Nature Genetics, 41*, 1272–1274.

Alders, M., Mendola, A., Adès, L., Al Gazali, L., Bellini, C., Dallapiccola, B., et al. (2013). Evaluation of clinical manifestations in patients with severe lymphedema with and without CCBE1 mutations. *Molecular Syndromology, 4*(3), 107–113.

Alitalo, K. (2011). The lymphatic vasculature in disease. *Nature Medicine, 17*, 1371–1380.

Bellini, C., Boccardo, F., Bonioli, E., & Campisi, C. (2006). Lymphodynamics in the fetus and newborn. *Lymphology, 39*, 110–117.

Bellini, C., Boccardo, F., Campisi, C., Villa, G., Taddei, G., Traggiai, C., et al. (2008). Lymphatic dysplasias in newborns and children: The role of lymphoscintigraphy. *J Pediatric, 152*, 587–589.

Bellini, C., Ergaz, Z., Boccardo, F., Bellini, T., Bonioli, E., & Ramenghi, L. A. (2013). Dynamics of pleural fluid effusion and chylothorax in the fetus and newborn: Role of the lymphatic system. *Lymphology*. in press.

Bellini, C., Ergaz, Z., Radicioni, M., Forner-Cordero, I., Witte, M., Perotti, G., et al. (2012). Congenital fetal and neonatal visceral chylous effusions: Neonatal chylothorax and chylous ascites revisited. A multicenter retrospective study. *Lymphology, 45*, 91–102.

Belov, S. T. (1993). Anatomopathological classification of congenital vascular defects. *Seminars in Vascular Surgery, 6*, 219–224.

Braamskamp, M. J., Dolman, K. M., & Tabbers, M. M. (2010). Clinical practice. Protein-losing enteropathy in children. *European Journal of Pediatrics, 169*, 1179–1185.

Brice, G., Child, A. H., Evans, A., Bell, R., Mansour, S., Burnand, K., et al. (2005). Milroy disease and the VEGFR-3 mutation phenotype. *Journal of Medical Genetics, 42*, 98–102.

Brice, G., Mansour, S., Bell, R., Collin, J. R., Child, A. H., Brady, A. F., et al. (2002). Analysis of the phenotypic abnormalities in lymphoedema-distichiasis syndrome in 74 patients with FOXC2 mutations or linkage to 16q24. *Journal of Medical Genetics, 39*, 478–483.

Child, A. H., Gordon, K. D., Sharpe, P., Brice, G., Ostergaard, P., Jeffery, S., et al. (2010). Lipedema: An inherited condition. *American Journal of Medical Genetics Part A, 152A*, 970–976.

Connell, F., Brice, G., Mansour, S., & Mortimer, P. (2009). Presentation of childhood lymphoedema. *Journal of Lymphoedema, 4*, 65–72.

Connell, F., Kalidas, K., Ostergaard, P., Brice, G., Homfray, T., Roberts, L., et al. (2010). Linkage and sequence analysis indicate that CCBE1 is mutated in recessively inherited generalised lymphatic dysplasia. *Human Genetics, 127*, 231–241.

Daniel-Spiegel, E., Ghalamkarpour, A., Spiegel, R., Weiner, E., Vikkula, M., Shalev, E., et al. (2005). Hydrops fetalis: An unusual prenatal presentation of hereditary congenital lymphedema. *Prenatal Diagnosis, 25*, 1015–1018.

Dubin, P. J., King, I. N., & Gallagher, P. G. (2000). Congenital chylothorax. *Current Opinion in Pediatrics, 12*, 505–509.

Enjolras, O., & Mulliken, J. B. (1997). Vascular tumors and vascular malformations (new issues). *Advances in Dermatology, 13*, 375–423.

Fang, J., Dagenais, S. L., Erickson, R. P., Arlt, M. F., Glynn, M. W., Gorski, J. L., et al. (2000). Mutations in FOXC2 (MFH-1), a forkhead family transcription factor, are responsible for the hereditary lymphedema-distichiasis syndrome. *American Journal of Human Genetics, 67*, 1382–1388.

Ferrell, R. E., Levinson, K. L., Esman, J. H., Kimak, M. A., Lawrence, E. C., Barmada, M. M., et al. (1998). Hereditary lymphoedema: Evidence for linkage and genetic heterogeneity. *Human Molecular Genetics, 7*, 2073–2078.

Fox, G. F., Challis, D., O'Brien, K. K., Kelly, E. N., & Ryan, G. (1998). Congenital chylothorax in siblings. *Acta Paediatrica, 87*, 1010–1012.

Gordon, K., Schulte, D., Brice, G., Simpson, M. A., Roukens, M. G., van Impel, A., et al. (2013). Mutation in vascular endothelial growth factor-C, a ligand for vascular endothelial growth factor receptor-3, is associated with autosomal dominant milroy-like primary lymphedema. *Circulation Research, 112*(6), 956–960.

Hennekam, R. C., Biesecker, L. G., Allanson, J. E., Hall, J. G., Opitz, J. M., Temple, I. K., et al. (2013). Elements of morphology: General terms for congenital anomalies. *American Journal of Medical Genetics*. [epub ahead of print].

Hennekam, R. C. M., Geerdink, R. A., Hamel, B. C., Hennekam, F. A., Kraus, P., Rammeloo, J. A., et al. (1989). Autosomal recessive intestinal lymphangiectasia and lymphoedema, with facial anomalies and mental retardation. *American Journal of Medical Genetics, 34*, 593–600.

Irrthum, A., Devriendt, K., Chitayat, D., Matthijs, G., Glade, C., Steijlen, P. M., et al. (2003). Mutations in the transcription factor gene SOX18 underlie recessive and dominant forms of hypotrichosis-lymphedema-telangiectasia. *American Journal of Human Genetics, 72*, 1470–1478.

Keeley, V. (2006). The use of lymphoscintigraphy in the management of chronic oedema. *Journal of Lymphoedema, 1*, 42–57.

Lee, B. B., Kim, Y. W., Seo, J. M., Hwang, J. H., Do, Y. S., Kim, D. I., et al. (2005). Current concepts in lymphatic malformation. *Vascular and Endovascular Surgery, 39*, 67–81.

Martinez-Corral, I., & Makinen, T. (2013). Regulation of lymphatic vascular morphogenesis: Implications for pathological (tumor) lymphangiogenesis. *Experimental Cell Research, S0014–4827*(13), 00034–00037. doi:10.1016/j.yexcr.2013.01.016 [Epub ahead of print].

Moffatt, C. J., Franks, P. J., Doherty, D. C., Williams, A. F., Badger, C., Jeffs, E., et al. (2003). Lymphoedema: An underestimated health problem. *Quarterly Journal of Medicine, 96*, 731–738.

Mortimer, P. S. (1995). Managing lymphoedema. *Clinical and Experimental Dermatology, 20*, 98–106.

Mulliken, J. B., & Glowacki, X. X. (1982). Classification of pediatric vascular lesions. *Journal of Plastic and Reconstructive Surgery, 69*, 412–422.

Rauen, K. A. (2013). The RASopathies. *Annual Review Genomics Human Genetics, 14*, 355–369.

Rezaie, T., Ghoroghchian, R., Bell, R., Brice, G., Hasan, A., Burnand, K., et al. (2008). Primary non-syndromic lymphoedema (Meige disease) is not caused by mutations in FOXC2. *European Journal of Human Genetics, 16*, 300–304.

Rockson, S. G., & Rivera, K. K. (2008). Estimating the population burden of lymphoedema. *Annals of the New York Academy of Sciences, 1131*, 147–154.

Smeltzer, D. M., Stickler, G. B., & Schirger, A. (1985). Primary lymphedema in children and adolescents: A follow-up study and review. *Pediatrics, 76*, 206–218.

Straats, B. A., Ellefson, R. D., Budahn, L. L., Dines, D. E., Prakash, U. B., & Offord, K. (1980). The lipoprotein profile of chylous and nonchylous pleural effusions. *Mayo Clinic Proceedings, 11*, 700–704.

Sutkowska, E., Gil, J., Stembalska, A., Hill-Bator, A., & Szuba, A. (2012). Novel mutation in the FOXC2 gene in three generations of a family with lymphoedema-distichiasis syndrome. *Gene, 498*, 96–99.

Tammela, T., & Alitalo, K. (2010). Lymphangiogenesis: Molecular mechanisms and future promise. *Cell, 140*, 460–476.
Tiwari, A., Myint, F., & Hamilton, G. (2006). Management of lower limb lymphoedema in the United Kingdom. *European Journal of Vascular and Endovascular Surgery, 31*, 311–315.
Van Balkom, I. D. C., Alders, M., Allanson, J., Bellini, C., Frank, U., De Jong, G., et al. (2002). Lymphoedema-lymphangiectasia-mental retardation (Hennekam) syndrome: A review. *American Journal of Medical Genetics, 112*, 412–421.
Vignes, S., & Bellanger, J. (2008). Primary intestinal lymphangiectasia (Waldmann's disease). *Orphanet Journal of Rare Diseases, 3*, 5.

Printed by Publishers' Graphics LLC
DBT140208.15.15.109